T0155769

AESS Interdisciplinary Environmental Studies and Sciences Series

Series editor
Wil Burns
Forum for Climate Engineering Assessment
School of International Service
American University
Washington, DC, USA

Environmental professionals and scholars need resources that can help them to resolve interdisciplinary issues intrinsic to environmental management, governance, and research. The AESS branded book series draws upon a range of disciplinary fields pertinent to addressing environmental issues, including the physical and biological sciences, social sciences, engineering, economics, sustainability planning, and public policy. The rising importance of the interdisciplinary approach is evident in the growth of interdisciplinary academic environmental programs, such Environmental Studies and Sciences (ES&S), and related 'sustainability studies.'

The growth of interdisciplinary environmental education and professions, however, has yet to be accompanied by the complementary development of a vigorous and relevant interdisciplinary environmental literature. This series addresses this by publishing books and monographs grounded in interdisciplinary approaches to issues. It supports teaching and experiential learning in ES&S and sustainability studies programs, as well as those engaged in professional environmental occupations in both public and private sectors.

The series is designed to foster development of publications with clear and creative integration of the physical and biological sciences with other disciplines in the quest to address serious environmental problems. We will seek to subject submitted manuscripts to rigorous peer review by academics and professionals who share our interdisciplinary perspectives. The series will also be managed by an Editorial board of national and internationally recognized environmental academics and practitioners from a broad array of environmentally relevant disciplines who also embrace an interdisciplinary orientation.

More information about this series at http://www.springer.com/series/13637

Kimberly K. Smith

Exploring Environmental Ethics

An Introduction

Kimberly K. Smith
Carleton College
Northfield, MN, USA

ISSN 2509-9787 ISSN 2509-9795 (electronic)
AESS Interdisciplinary Environmental Studies and Sciences Series
ISBN 978-3-030-08447-9 ISBN 978-3-319-77395-7 (eBook)
https://doi.org/10.1007/978-3-319-77395-7

Softcover re-print of the Hardcover 1st edition 2018

Printed on acid-free paper

This Springer imprint is published by the registered company Springer Science+Business Media, LLC part of Springer Nature.
The registered company address is: 233 Spring Street, New York, NY 10013, U.S.A.

Preface

Exploring Environmental Ethics is just that: an exploration. It is not a typical scholarly monograph offering an ethical theory, nor is it a stand-alone textbook intended to serve as a comprehensive introduction to the field of environmental ethics for students of philosophy. It is instead an introduction to the discipline of environmental ethics for people outside of that field. Specifically, it is intended for students in interdisciplinary environmental studies programs who want to engage with the scholarly literature on environmental ethics but don't have training in philosophy.

I wrote this text to serve what I see as a growing need. As interdisciplinary environmental studies programs become more common in higher education, we need teaching materials that not only expose students to typical scholarship in different academic disciplines but also help *explain* different academic disciplines. We need texts that clarify disciplinary assumptions and methodologies, highlight important cross-disciplinary conversations, and discuss how terms may acquire different meanings as they migrate across fields. In short, we need texts that help us navigate the often confusing interdisciplinary landscape of environmental studies.

I come to this project not as a professional ethicist but as an intellectual historian whose primary field is political theory. From this vantage, standing just outside the field of ethics, I explore conversations among scholars of environmental ethics that are central to environmental studies as a field. Drawing on my experience teaching environmental ethics, I also discuss topics that are particularly important to environmental studies students but not typically addressed in standards texts on environmental ethics. My hope is that the book will serve as a map or guide, a resource that can help readers confidently to pursue their own explorations of this rich and important field.

In developing this book, I benefitted from conversations with and feedback from many people, including Dale Jamieson, Jennifer Everett, Jeremy Bendik-Keymer, the members of the 2009 NEH Workshop on Aldo Leopold, Glenn Adelson, Daniel Groll, Roger Jackson, Thabiti Willis, the anonymous reviewers at Springer, and the many students in my Environmental Ethics courses. I am particularly grateful for the many conversations I have had with my mother, Judy Smith, who is the inspiration for the character of Judy K. I have tried to capture some of her wisdom about land management, stewardship, and the good life in this book.

Northfield, MN, USA Kimberly K. Smith

Contents

Chapter 1
Introduction

Contents

1.1 Managing Spring Lakes

Judy K. owns a 70-acre piece of land in rural mid-Michigan that she has named "Spring Lakes." The land includes two small ponds fed by a chain of small streams. It also has a couple of marshes and a forested "mountain" (more properly, a large hill). There's evidence of a couple of Indian burial mounds on the land as well, left by the Potawatomi people who once lived in this area. Judy lives on this property and also rents out several small cottages to tenants.

As she manages this land, Judy has to make several kinds of decisions: Should deer hunting and muskrat trapping be allowed? Would it be a good idea to cut down some diseased oaks on the north side of the mountain? Should she prevent the ponds from filling with algae and weeds to make them suitable for swimming, and if so, should she use chemicals, periodic dredging, or some other strategy? Should she permit fishing, and if so, under what rules? Should she take measures to discourage Canada geese from making their home in the ponds and fouling the lawns, or abandon the attempt to maintain green, park-like lawns around the ponds? What about the Potawatomi burial mounds — does she have a duty to inform the local historical society about them? What if removing the diseased oaks would disturb the mounds?

All of these decisions require Judy to make environmental value judgments: judgments about how and how much we should value various features of the nonhuman environment: individual plants and animals, species, ecosystems, and all their varied interactions. But to complicate matters, Judy doesn't make these judgments in isolation. She often hears opinions (solicited and unsolicited) from her tenants, her children, and her neighbors, all of whom feel they have a stake in the property.

K. K. Smith, *Exploring Environmental Ethics*,
AESS Interdisciplinary Environmental Studies and Sciences Series,
https://doi.org/10.1007/978-3-319-77395-7_1

State officers from the wildlife, public health, and environmental quality agencies also frequently visit with Judy to discuss their concerns and advise her about the applicable environmental regulations. Whose interests should Judy consider in making these decisions? What should her land management goals be? How far should the state dictate what she does with this land?

1.2 What This Book Is About

Not all of us are lucky enough to own a 70-acre piece of land in mid-Michigan, but like Judy K., we all face questions about environmental values. Whether you're a policy professional, an industry executive, a student, or a homemaker, you will be faced with decisions that affect nonhuman nature. You might have to make decisions about how to comply with environmental regulations or how to dispose of your plastic water bottle. You might have to decide what to plant in your garden or whether to use chemicals on your lawn. And you might also have to decide whether to support political candidates and causes seeking to impose or change environmental laws. This book is about what to think about when you're making those decisions. More precisely, it's about how the scholarly literature on environmental ethics can help you make those decisions.

Making these decisions about environmental values involves, in part, deciding what moral principles or rules you wish to live by — that is, what *duties* you owe to other humans and nonhuman nature. This is a central question in the field of environmental ethics. But environmental ethics includes much more than following rules. It also involves thinking creatively and sensitively about how your relationship to nature contributes to a meaningful, choiceworthy life. That is, environmental ethics is about how to live *the good life* in relationship to nature and in relationship to other beings who also depend on the natural world. In sum, environmental ethics is about what your relationship to the nonhuman world means to you: What role it plays in your vision of a good life, what duties you have toward human and nonhuman beings with regard to natural resources, and whether you see yourself as a steward of the earth.

This text is intended to help you think through those questions. It is designed for courses in environmental ethics that are integral to environmental studies and sciences programs. I envision the reader as a student (at the undergraduate or graduate level) who expects environmental stewardship to be an important part of his or her life, as a citizen, a policy maker, or an environmental management professional. I also envision the reader as someone taking a course in an American college or university; the examples are therefore mostly from the United States, and the textbook as a whole has an "American accent," so to speak. However, many (perhaps most) environmental ethics scholars are not from the United States. This is very much a multi-national field, and I will be drawing on some important scholarship from people living and working in other countries.

1.3 Ethics as Part of an Interdisciplinary Environmental Education

This text is written in particular for people interested in an interdisciplinary approach to environmental issues. Many colleges offer discipline-based courses on environmental affairs: One could study chemistry, ecology, physics, geology, political science, economics, history, or philosophy, and apply the theories and methods learned in any of those fields to environmental problems. That approach will give you some useful tools and valuable insights. But interdisciplinary programs are based on the conviction that environmental problems are rooted in *interdependent social-ecological systems*, and no single discipline can adequately explain how those systems work or how best to manage them. We need insights from many disciplines, and we need effective ways to integrate what we learn in these disparate disciplines.

That's difficult, because every discipline is based on different and even incompatible assumptions: assumptions about what nature is, what causal factors we should focus on, what is the appropriate time scale and geographic scale for making sense of the world, and even what counts as "truth." An ecologist, for example, usually studies landscape-level phenomena (that's the geographic scale) and considers changes that occur over a period of a few years at best (that's the temporal scale). The ecologist thinks of nature as a dynamic, complex system, in which plants, animals, climate, hydrology, and other nonhuman factors interact to cause the phenomena we observe. And she verifies her conclusions by collecting data from her study area and using statistical methods to draw inferences to larger phenomena and larger areas. An economist, in contrast, might focus on national or international economic systems, examining changes taking place over a few decades. She assumes that human choices (reflected in public policy or consumer behavior, for example) cause the phenomena we observe. She also collects data and draw inferences—that's an important similarity—but she probably thinks of nature (if at all) as a storebox of resources, not a dynamic system. Little wonder the ecologists and economists often have difficulty talking to each other about environmental issues!

Ethics is a branch of moral philosophy, which is also an academic discipline based on certain assumptions. Ethicists have traditionally focused on how choices made by individual human beings affect other human beings in the near term, rather than on how complex natural or social systems operate over several decades or centuries. But environmental problems in an age of global climate change result from the complex interactions of thousands or millions of small decisions and affect people living halfway across the world or in the distant future. The traditional disciplinary approach to ethics has had to evolve to deal with such issues: Environmental ethicists have had to expand the *geographic* and *temporal scale* of their thinking. They have had to engage with the natural sciences and think about whether we have moral duties to things like species or ecosystems. And they have had confront the fact that in the real world, individuals may not have much choice about how they interact with the environment, since they're trapped in complex social-ecological systems that limit their options.

This text is part of an effort to bring the discipline of ethics into deeper conversation with other disciplines that contribute to managing environmental problems. It will invite you to think about how an ethicist's perspective differs from the perspectives you encounter in other environmental studies courses. For example, ecologists have standard definitions of concepts like "ecosystem" and "biodiversity." Ethicists often question those definitions, asking what values they are (implicitly) based on. Similarly, economists have a strategy for deciding how much value to put on future benefits (called "discounting"). Ethicists have questioned how and why economists tend to discount those future benefits. Is it philosophically and ethically justified to value benefits to future generations differently than we value the benefits we enjoy today? Becoming sensitive to differences between disciplines — where different scholars are "coming from", what assumptions they're making, and how their research relates to work being done in other disciplines — is very important for anyone involved in interdisciplinary work. I hope this text will help you develop that sensitivity.

1.4 Overview of the Book

This text doesn't try to cover every topic in environmental ethics in detail. Instead, it selects several important topics and provides a general overview of the scholarly discussion among ethicists (and between ethicists and other disciplines) concerning that topic. This overview is meant as a guide to the scholarly literature in this field (but not as a substitute for reading that literature yourself!)

Chapter 2 discusses the nature of ethical inquiry and why we need an ethic that explicitly addresses our relationship to the natural world.

Chapter 3 introduces various concepts of justice that can inform our thinking about environmental policy. It also discusses how far the government should go to protect the environment (that is, what are the proper limits of government power?) and what we can expect of citizens and corporations in relation to the environment (that is, what are our civic obligations?)

Chapters 4 and 5 consider to whom we owe duties of justice, in addition to the humans we live with: Do we have duties to animals, plants, species, ecosystems? What about future generations? How do they figure into our decision making?

Chapter 6 explores the meaning and value of property rights and how stewardship can contribute to a good, choiceworthy life.

Chapter 7 takes up another central issue in environmental policy: What values or meaning do we find in the landscape? What makes a landscape worth preserving? How does interpreting landscapes relate to justice and the good life?

Finally, Chapter 8 considers the ways one can make room for environmental stewardship in one's own life.

This short text is merely a beginning, an attempt to give you a foundation for further reflection and exploration. Each chapter ends with suggestions for further readings in the field of environmental ethics. Ultimately, however, developing an environmental ethic is a matter of practice—of developing habits of thoughtfulness, attentiveness, and care in your dealings with the nonhuman world.

Further Reading

Orr, D.: Ecological Literacy: Education and the Transition to a Postmodern World. SUNY Press, Albany (1992)

Soulé, M., Press, D.: What is environmental studies? Bioscience. **48**(5), 397–405 (1998)

Chapter 2
Why Study Environmental Ethics?

Contents

2.1 The Problem of Purple Loosestrife

Judy K. was enjoying a walk along the pond when she noticed an odd sight. Maria, her neighbor, was wading along the bank of the pond and pulling up the bright purple weeds growing in the shallow water. Maria waved her harvest and called out, "Purple loosestrife! You have to get rid of it!" Judy, confused, asked why. "Purple loosestrife is an invasive species around here," Maria told her. "It crowds out the native plants. If you let it spread, it can change the lake ecosystem. It might also harm wildlife by eliminating natural foods and habitat." Judy was perplexed. The plant didn't seem to be taking over, although she'd seen other nearby ponds with much larger and denser concentrations of loosestrife. "But I like purple loosestrife," she objected. "And why should I prefer native plants to "invasive" ones?" "Don't you care about biodiversity?" Maria asked. Judy pondered the question. She had never really thought about biodiversity. Should she care about it? Did she have some

K. K. Smith, *Exploring Environmental Ethics*,
AESS Interdisciplinary Environmental Studies and Sciences Series,
https://doi.org/10.1007/978-3-319-77395-7_2

sort of duty to get rid of purple loosestrife, even though she hadn't planted it there in the first place? She now felt vaguely uneasy about the purple loosestrife, but she wasn't sure whether she really wanted it gone.

2.2 Why Ethics?

Judy K. is encountering an ethical dilemma of the sort most of us face quite often. She thinks she may have a moral obligation to do something. Her neighbor certainly thinks so. But Judy herself isn't so sure. She needs to think through whether uprooting the purple loosestrife would be a good choice, as well as how she will explain her decision to her friend Maria. The study of ethics is meant to help people in Judy's situation.

Ethics is a field in philosophy focuses on explaining and defending judgments about right and wrong conduct. We often think of ethics as being about what rules you should follow in order to be a good person. But we can also define ethics more broadly as the study of what contributes to a good or choiceworthy human life. That is, it's not only about rules but also about what goals are good to pursue, how to live well and richly in the world. Your relationship to the natural world can contribute to a good life in many ways. Most obviously, we all rely on nature for the food we eat, the clothes we wear, and all the things we use in our lives. The quality of the natural world affects our health and wellness. Using natural resources also affects our relationships to other humans — the people who produced, sold, needed, or bought these things. So our relationship to nature is part of social ethics.

But we don't just use nature; we also appreciate it. We are awed by its grandeur, humbled by its vastness, soothed by its beauty, and delighted by its unexpected quirkiness. We test our fitness and discipline against mountains, measure our fortitude against northern winters, and discover the depths of our compassion when faced with animal suffering. We may take our artistic or spiritual inspiration from the natural world. Or, on the other hand, we may turn away from the beauty of nature in order to seek spiritual enlightenment. Ethics concerns the value of these experiences as well.

Finally, you may be persuaded that you have a duty to respect living organisms simply because they (like yourself) are alive; they are capable of flourishing in their characteristic way, and that deserves respect. In other words, you may embrace an ethic that shows respect not just for humans but for nature itself.

Most people want to live well. They also want, at the very least, to be able to justify their choices — to explain why they live the way they do. For example, we want to think clearly about what sort of experiences with the natural world we value and why. We want to think clearly about whether our choices (about where to live, what job to pursue, what food to eat, what consumer goods to buy, etc.) are leading us into the kind of relationship with nature that we desire. And we want to know that we are satisfying our moral obligations. The study of ethics should help us achieve that clarity.

2.3 The Method of Ethical Inquiry

But how can we explain the reasons for choices that often seem so deep, complex, and ultimately personal and idiosyncratic? How can we hope to bring reason and order to these decisions?

Philosophers have been confronting this question for thousands of years. Some of our most important works in ethics come from a civilization on the Greek peninsula that flourished about five thousand years ago. This civilization gave us not only Plato and Aristotle but the foundations of what the contemporary world calls the Western tradition of moral philosophy. This text is part of that tradition. The Western tradition of moral philosophy is only one of many conversations about the good life going on around the world; in fact, it's only one of many conversations within Western societies themselves. For example, religious practices typically also include a good deal of ethical reflection and reasoning, and there is a lot of cross-disciplinary discussion between scholars in religious studies and in philosophy.

But moral philosophers usually try to reason about moral issues in a way that is independent of any particular religious tradition. That is, instead of beginning with the authority of a religious text or precept, we begin with our own moral intuitions or beliefs. We then subject those beliefs to *rational critical analysis*. That is, we try to determine whether some of our beliefs conflict with others that are better justified. This may involve working back and forth between specific judgments and broader principles, to make sure they are consistent. Philosophers typically call this back-and-forth process the method of *reflective equilibrium*. The goal is to achieve internal consistency among all our moral beliefs. We may find that some of our beliefs have to be revised, or even abandoned altogether.

For example, we might take up a question that we know is controversial or contested, like: Is it wrong to kill animals for food? We approach that question by beginning with some judgment that we think is much less controversial, like: It is wrong to kill humans for food. We can then reason from that shared commitment: If we're right about not killing humans for food, what else must we hold to be true? We could begin by articulating why it's wrong to eat humans: is it something about humans, some characteristic they have that makes them non-edible? We might also want to consider cases in which humans have become food: what was it about those circumstances that made eating humans seem acceptable? We might end up qualifying our judgment: Perhaps it's only wrong to eat humans when there is other food available, for example. We would then consider whether animals and humans are different in any morally important respect. Along the way, we would also have to call on what we know about animals and humans. (Do they feel pain, in the same ways and to the same extent, as humans? Do they know they are destined to become food? Do they care?) At the end of this process you might conclude there are good reasons for thinking animals are different from humans in ways that make it acceptable to eat animals. Of course, I might decide otherwise. Ethical analysis doesn't promise to end the debate on difficult ethical questions. But the process should give you a clearer idea of the reasons for your position. Those are reasons that I will probably have to respect, even if I don't agree with them. In other words, your moral position will be well-justified, and you will be much clearer and more certain about what that position is and why you hold it.

Perhaps you're wondering, "Why bother with all this? I feel that what I believe is right; isn't that enough?" It's certainly true that you don't need to study ethics in order to be a good person. Virtues are largely a matter of habit and good socialization. But the study of ethics is meant to help people reflect critically on the values and habits they've acquired, uncritically, from their upbringing. This is particularly important in a culture shaped by individualism and consumerism, which can promote environmentally irresponsible practices. What you *feel* is right may, in fact, be very wrong—and critical examination can help us to uncover and correct those habits and values.

2.3.1 Ethical Inquiry Versus Scientific Inquiry

Notice how different the method described in this section is from the methods used to discover truth in natural or social science courses. Ethicists don't normally conduct experiments or collect data to test their theories. On the other hand, like scientists, they do try to be crystal clear about what their assumptions are, and they focus a great deal on whether the inferences we draw from our assumptions are logical and coherent.

As discussed in Chapter One, ethicists typically make some assumptions about the *geographic* and *temporal scale* of the moral community. For example, an ethical analysis might assume that the community is small enough that you can easily find out how your actions will affect others. It might assume that it's easy to know which actions caused which results. These are *simplifying assumptions* aimed at making a problem more manageable, more suitable for ethical analysis. But in our global, multicultural, environmental age, those assumptions don't hold up well. We know our actions have impacts beyond our immediate community and beyond our lifespan. We know that we are part of complicated systems, and that our own actions can play a small role in causing great harm.

One consequence of our contemporary ethical situation is that we see the need for large-scale, long-term strategies for managing environmental affairs — strategies that require global cooperation. Therefore, we feel a lot of pressure to find out whether our moral intuitions about environmental issues are shared by others, especially people from distant cultures. That leads us to other disciplines, like sociology and anthropology, and to the idea of cultural relativism, discussed in Sect. 2.4.2.

A Buddhist Approach to Ethics

Other ethical and faith traditions also engage in moral reasoning. Here is an example of Buddhist ethical reasoning on the subject of eating animals, courtesy of Carleton College Professor of Religion Roger Jackson:

"The reason meat-eating is a largely unsettled question for Buddhists is rooted in the tension between two fundamental guidelines established (so we are told) by the Buddha.

"The first is the famous "First Precept," incumbent upon all monks, nuns, and avowed laypeople, which enjoins Buddhists to refrain from taking life. The precept itself does not specify which forms of life, but it is generally taken to cover all "sentient beings," i.e., anything from humans "down" to insects but not including plants. In the fine-tuned calibration of karmic consequences referred to as the "heaviness" of karma, it is clear that killing a human is worse than killing an animal, the usual rationale being that a human rebirth is richer in spiritual potential than an animal birth. Still, to kill any animal is to incur negative karma, hence to risk negative consequences in this life and in future lives.

"The second guideline is the specific instruction the Buddha gave to his followers when asked whether it was acceptable to eat meat. He replied that as long as the meat fulfilled three criteria, it was acceptable to eat it. The three are: (a) you did not kill it yourself, (b) you did not order it killed for you, and (c) to the best of your knowledge, it was not killed expressly for you. In the Buddhist moral calculus, intention is the key to moral responsibility, and fulfillment of these three conditions, for most Buddhists, exculpates the meat-eater. (It's important to note that Buddhist monks and nuns are, literally, "beggars" (bhikkhus/bhikkhunis), and so were expected to eat whatever food was put into their bowls as they made their alms-rounds. This may be the practical concern governing the allowance of meat-eating, and it certainly is the stipulation that has allowed even those Buddhists who are scrupulous about the precepts to eat meat in good conscience.)

"This line of reasoning obviously begs a number of questions, e.g., about the degree to which we are complicit in killing when we buy meat at the supermarket. Indeed, it was persistent student arguments to this effect that persuaded me to become vegetarian for good about twenty years ago, but the reasoning they – and I – used is an extension of Buddhist principles rather than a strict reading of them, and clearly implies that the First Precept trumps any qualifications that may have been added to it, even by the Buddha himself. (Roger Jackson, email letter to author, June 9, 2012)."

2.3.2 Empirical Versus Normative Study of Values

Moral philosophy is not the only academic discipline that is interested in studying values. For example, biologists sometimes study ethics as well. Some biologists treat ethics as a behavior that evolved because it has survival value for human communities: Ethical communities (under this reasoning) are more likely to thrive and pass on their genes. (Aldo Leopold uses a version of this argument in *A Sand County Almanac* (1949), which we will discuss below in Sect. 2.5) Ethicists don't necessarily disagree with this evolutionary account of why humans have the capacity for ethical reasoning. But they insist that ethics can't be reduced to biology. Once the capacity for ethical reasoning evolved, ethical systems took on a life of their own; ethical rules are *not* simply whatever rules help the species to survive. Some behaviors motivated by ethical values may not have survival value at all, and not all behaviors that do have survival value are actually ethical. For an ethicist, explaining the survival value of a practice will never answer the question of whether the practice is ethically justified.

Most of the social sciences also have some methods and theories for investigating values. But social scientists typically take an *empirical* approach to the study of ethics and values. That is, they are interested in describing what people believe and why. In contrast, the kind of ethical analysis we described above is called *normative* analysis: It focuses not merely on describing values (although accurate descriptions help!) but on evaluating whether those values are consistent and well-justified.

For example, let's compare ethics with a field that also considers the study of values to be its domain: Cultural anthropology. Anthropologists typically study moral values as a part of a specific culture they are interested in understanding. They don't ask whether those values are ultimately true or false, right or wrong. Rather, their goal is to understand what those values are and how they fit into the overall worldview of members of that culture. To do this sort of investigation, a good anthropologist learns to set her own ethical values aside — to suspend judgment about the morality of social practices — so that she can enter into the perspective of the people she is studying.

This is where anthropology and ethics differ. Indeed, the anthropologist's open-minded, nonjudgmental stance toward ethical values sometimes seems to be completely at odds with the sort of normative ethical analysis described above. To the ethicist, anthropologists may seem too uncritical of the values they are studying. To the anthropologist, ethicists may seem too disrespectful of the people who hold the values being evaluated. But these disciplinary differences don't have to lead to such conflict. On the contrary! We just have to keep in mind what each approach is trying to achieve. Normative ethical inquiry can help you to clarify what you value and why, so you can explain yourself to others. Or perhaps it's better to say that it clarifies the outer bounds of what you *can* believe, identifying the point at which you have to say, "Well, I guess I really don't understand how someone could believe *that*." And that's exactly where the anthropologist can step in. Anthropological investigation can help you to understand ethical perspectives that are radically different from your own. Even better, it can explain how your own ethical perspective looks to people who aren't familiar with your culture. In short, both the anthropologist and the ethicist can contribute to better understanding of and communication about values. In our age of global environmental problems, that kind of cross-cultural understanding is particularly important.

Animal Sacrifice

An example of moral (and political) conflict rooted in cultural differences is the controversy over animal sacrifice. Animal sacrifice is an important part of the Caribbean religion Santeria. Some animal rights activists have called for an end to this practice, characterizing it as an example of human arrogance and dominance over animals. But for practitioners of Santeria, sacrifice is the essence of humans' relationship to the divine. Prayers merely express what people want from their relationship with the gods; sacrifice, as an act of communion, actually constitutes that relationship, making real the reciprocal bond between humans and the gods. Indeed, far from expressing human dominance of nature, the sacrifice recognizes the physical and spiritual *interdependence* of living things. As scholar Joseph Murphy explains, "Animals die so that human beings may live… all are related by delicate exchanges and balances of nature that make human life possible. …. [Animal] blood is offered to the *orishas* to show human beings their dependence on the world outside them and to give back to the invisible world something of what it gives to the visible" (Murphy 1988, p. 44). The practice is grounded on a view of the entire material world as sacred, as an expression of spiritual reality whose wholeness depends on making such exchanges. Animal sacrifice expresses humans' dependence on and unity with the whole material and spiritual world.

Animals are particularly powerful and important in this worldview; even the *orishas* depend on the vitality of animal life. Some practitioners insist that the animal offers itself during the ritual, just as a human possessed by a spirit may be said to offer or sacrifice himself to that spirit. But whether or not the animal can be seen as a voluntary participant, the sacrifice itself is meant to be an act of profound humility and gratitude on the part of humans for this gift of life.

Understanding this perspective may allow practitioners of Santeria and animal rights activists to understand one another. Perhaps they can even find common ground and work together to address threats to animal welfare.

Nevertheless, the more we learn about cultural differences in moral values, the more we may wonder, "Is there really a universal right or true answer to ethical questions? If not, what is the point of ethical analysis?" To explore those questions, we need to take a brief tour of metaethics.

2.4 A Brief Tour of Metaethics

The field of ethics is usually divided into two subfields: normative ethics and metaethics. Metaethics takes up big, abstract questions like what does it mean to say that it's wrong to do something? Are there absolute moral truths that we are trying to discover, as we try to discover the laws of physics? Or do we mean something different by "moral truth" than we do by "scientific truth"? This book is mostly concerned with normative and applied ethics; that is, it focuses on what counts as a good reason for a moral judgment or what we should do in a particular situation. So we aren't going to delve too deeply into metaethics. But there are some useful ideas that we can borrow from metaethics to orient us as we delve into more specific questions about how we should handle an ethical dilemma.

2.4.1 Moral Objectivism

Some philosophers argue that moral statements (like, "It is wrong to cause nonhuman animals unnecessary suffering") should be taken at face value: They are making a claim about objective reality, and they can be true or false.

There are different ways to make this argument. One might argue that this statement is true in the same way that a proposition like 2 + 2 = 4 is true, that it just follows logically from how we define our terms. This view is supported by the fact that some moral disagreements can be worked out by being clearer about what we mean. But not all moral debates can be resolved this way; we often seem to be arguing about the way the world is, not just about what we mean.

Another way to make the argument is to say that moral principles are facts about the way the world is — but they aren't "natural" facts, like facts about the specific gravity of water. That is, we can't do some research, gather some evidence, and figure out what's true and false in this domain, as we do when we disagree about physical phenomena. But that doesn't mean there isn't a truth of the matter. Humans may simply have limited abilities to know or prove some truths to everyone's complete satisfaction. Perhaps there is a super- or non-natural realm where moral truths reside. That would be consistent with many faith traditions that teach that there is a God, or gods, or a spiritual reality that is concerned with human morality. But one could accept this position even if one doesn't believe in such spiritual realities, on the grounds that scientific methods are limited to dealing with only some kinds of truth. Someone who embraces this position then has some work to do to explain how we can discover these truths.

There is another possibility, however. Perhaps moral claims can be objectively true because they are rooted in human nature. Moral statements, under this view, are claims about how creatures like us should live, if we want to thrive. If human nature and social life are sufficiently uniform and stable across cultures, then human nature could provide an objective foundation for morality. What's true for the Hottentot should also be true for the American college student. (Note that this position seems like it could be verified through empirical research, and tested against the findings of cultural anthropology.)

2.4.2 Moral Subjectivism

Moral subjectivists have a hard time believing that moral principles are objectively real — that they are "out there" in the world — in the same way that scientific facts are. Instead, they suggest that moral claims express a feeling or attitude of the speaker. That is, the claim "It is wrong to cause nonhuman animals unnecessary suffering" expresses the speaker's disapproval of that practice.

We might be tempted to object to subjectivism on the grounds that it means people can just believe anything they choose — that they can choose to believe that murdering people in cold blood is OK, for example. But a serious subjectivist doesn't have to believe that you can believe anything you want. If your moral values express your true, deeply-held attitudes, those values probably aren't under your control. Someone who insists that murder feels just fine is probably lying, and probably feeling guilty about the lie, too.

There are different versions of subjectivism, too. One version is called *cultural relativism*. Here we have to be cautious, however, because "cultural relativism" can have different meanings in different disciplines. (Some of these meanings are explained in the box below.) Here I am talking about the view that your attitudes of moral approval or disapproval only make sense within your own cultural context; there is no moral standard that transcends the culture that generated it. Therefore, when we evaluate the truth of a moral statement, we are saying that it is more or less true *according to the standards of culture X*. This means that what is true for members of one culture is not necessarily true for members of another culture. Of course, different cultures might share the same standard, and cultures may influence one another, so this position doesn't necessarily mean that members of different cultures must always disagree. But it does means that moral judgments can have only *intersubjective validity*. That is, they may be valid for the community or communities with which we are reasoning, but not for all humans in all times and places.

Not all subjectivists are cultural relativists, however. One could argue that subjective attitudes are rooted in something deeper than culture, in our basic emotional make-up or psychology. Moral claims, under this view, would still express attitudes, but those attitudes might be universal—or idiosyncratic but stubbornly resistant to change.

Subjectivism is appealing because it recognizes the important role that culture and individual psychology plays in shaping our moral perspectives (and it may help to explain why moral disagreements are so hard to resolve). But it doesn't account for the role of reasons in moral judgment. We would like to say that there is a difference between an unexamined feeling of disgust at seeing an animal killed for food and a carefully reasoned argument that such practices are wrong. We wouldn't expect everyone to share our disgust reaction, but we probably do expect others to be persuaded by our reasons. We're inclined to say that a feeling or attitude that isn't supported by good reasons probably isn't a good basis for a moral judgment. Moreover, we believe that many people, and even people from different cultures, can reason together about what is morally right and good. All of this is hard to reconcile with the belief that moral judgments simply express subjective attitudes about the world.

Cultural Relativism

"Cultural relativism" is a term used in many disciplines, and it has taken on many different meanings. For example, in anthropology the term can refer to at least four different ideas:

Descriptive cultural relativism: The idea that social and psychological facts about humans are determined by their culture, so explanations of human variation should refer to cultural differences. There are different views about the scope and depth of this cultural determinism; its strongest version leads to epistemological relativism.

Epistemological cultural relativism: This position insists that the human mind (the very concepts we use to think with, perhaps even logic itself) varies by culture. Therefore, one can understand social phenomena only from within a culture. In its strongest form, this position casts doubt on the possibility of cross-cultural understanding or social theories that transcend specific cultures.

Normative cultural relativism: This position is also based on descriptive relativism, but it needn't embrace epistemological relativism. Normative relativism holds that all moral standards (like other social facts) are determined by culture. There are no trans-cultural standards we could refer to in order to evaluate whether moral judgments are true or false. This is the kind of relativism that moral philosophers are usually talking about, and the kind discussed above.

Finally, there is methodological cultural relativism: This refers to the stance taken by anthropologists when they are studying a culture. They will temporarily suspend their moral judgments and attempt to enter into the point of view of their subjects. More generally, this term can refer to the general orientation of anthropologists toward their subjects: a sensitivity to cultural difference and a commitment to respecting and valuing a culture's own understanding of itself.

2.4.3 Pragmatism

Some philosophers try to sidestep the problems on both sides of the objectivism/subjectivism debate by arguing that we should think of moral principles not as expressing the speaker's attitude or making truth claims about the world "out there."

Rather, these principles are *tools* we use to navigate the moral world. A good moral principle, like a good tool, isn't "true." Rather, it is a useful guide to a moral situation. Instead of worrying about whether moral statements are true, we should worry about how well they work.

Some people object to pragmatism on the same grounds that they object to subjectivism — that is, they fear it means that people can just choose any moral position at all, because "it works for me." More generally, it doesn't seem to allow for moral claims to be universalized. But pragmatists could respond that what works for you isn't a matter of your own free choice, and it probably isn't idiosyncratic. The world we're trying to navigate with these moral claims has an objective reality; that's why some moral principles seem to work well for virtually everyone. We are all humans trying to operate in a shared physical and social environment. It's not surprising that we all tend to hit on the same basic set of moral principles. (You can try this out yourself: Make up some weird moral system and try to live by it for a day. See how well it works.)

A deeper objection to pragmatism is that if we're going to evaluate a moral principle by how well it works, we have to define what we mean by "works." If we mean that this principle helps people to live a good, flourishing, happy life, then we seem to be back to the objectivist position that moral principles are rooted in human nature; they are claims about how creatures like us should live, if we want to thrive. On the other hand, if one believes that what "works" is defined by one's culture or individual psychology, then we seem to be back to subjectivism. Still, even if pragmatism can't entirely avoid these metaethical questions, it does remind us that our theoretical debates about values should not lose sight of their practical implications. For example, an influential group of environmental ethicists have been inspired by pragmatism to conduct empirical research on how environmental values work in practice.

In sum, I am inclined to think that each metaethical position captures some important dimensions of moral reasoning. Subjectivists are correct that moral statements express approval and disapproval, and cultural relativists remind us that those statements might be deeply shaped by our culture or individual psychology. But objectivists are right that moral claims attempt to say something about the world, not just about our feelings or attitudes. Moral judgments are based on reasons that at least strive to be universal. And pragmatists are right to point out that moral claims must not only make sense in theory; they should also be useful guides for action.

2.4.4 Purple Loosestrife Revisited

What does this discussion of ethical theory mean for Judy K.'s purple loosestrife situation? Unfortunately, none of what we've investigated so far will tell Judy what to do about the plant. But the discussion does suggest some lessons to keep in mind as she grapples with the question.

- The *subjectivist* might tell Judy that although her values seem true to her, they may ultimately rest on subjective intuitions that have (and can have) no objective truth value. So she should be careful not to assume that anyone who disagrees with her is just being unreasonable.
- The *cultural relativist* would suggest that her ideas about purple loosestrife are probably rooted in deep cultural beliefs about the value and meaning of the landscape, and it might not be possible to transcend those beliefs in making judgments. At the very least, the relativist position reminds us that it would be useful for Judy to delve deeper into why Maria sees the plant as an undesirable invader, and to explore where her own views of loosestrife came from.
- But even a cultural relativist may acknowledge that such cultural differences can at least be understood, and if they aren't too deep, Judy and her neighbors may reach an *intersubjective consensus* on how to value the purple plant.
- The *pragmatist* might add that in working toward this consensus, Judy and her neighbors should focus on finding an ethical approach that works in practice.
- The *moral objectivist*, however, would point out that ultimately we want our values to be not just useful but *true*, in an objective sense. The objectivist would urge Judy to keep questioning her own and her neighbors' views about invasive species and pond management. The process of rational-critical inquiry should help Judy's views to keep evolving. Ultimately she should develop a well-justified position, resting on principles that (if not universal) at least appeal far beyond her small community.

In sum, Maria's challenge to Judy could be the beginning of a valuable exercise in ethical reasoning. Even if it turns out that Judy has little ability to do anything about the spread of purple loosestrife, at least she will know what she thinks about that issue and why she thinks it. This is an important step toward living a good life. But will this process of ethical reasoning improve Judy's relationship to nature? We have considered why it's useful to study ethics, but why do we need *environmental* ethics?

2.5 Why Environmental Ethics?

The study of ethics is ancient; the academic field of environmental ethics is relatively new. Aldo Leopold's *A Sand County Almanac* (published in 1949) was one of the first major works that called for a new ethic toward nonhuman nature. Among those who heard that call was Richard Sylvan (né Routley), whose article "Is There A Need for a New, an Environmental, Ethic?" is a seminal paper in the academic study of environmental ethics. Leopold and Sylvan made a similar point: Until recently, ethics focused on duties of humans towards other humans. But many of us have a sense that we have moral obligations towards nonhumans—towards animals, for example. Sylvan captures this notion quite well in his *"last man" scenario*: Imagine the last man surviving the collapse of civilization and the extinction of all other human life on earth. Would it be wrong of him to destroy the remaining plants and animals? If there are to be no more humans left to value those nonhuman entities, who would he be wronging by his destructive actions? Many of us sense that there would be something morally objectionable about those actions (that is, this is a moral intuition that seems to be widely shared). There are a number of ways to explain why his actions might be wrong: Perhaps the nonhumans in question have moral value independent of any human values, or perhaps we see in his actions a character defect that makes his life seem less choiceworthy. There are other ways to explain our negative reaction to his actions. But most explanations suggest that we believe harming nonhumans is morally wrong even if we aren't violating any duties to other humans. Sylvan argues that this intuition suggests the need for a new ethic, one that recognizes the value of nonhuman nature and our duties toward it.

Leopold's argument also draws our attention to moral duties that extend beyond the human community. His essay "The Land Ethic" uses a form of reasoning called *moral extensionism.* He asks us to imagine that we are members not only of a human community but of a *biotic community*: the collection of plants, animals, and their interconnections that form the ecological context within which humans live and on which they depend. This exercise in moral imagination makes it easier to conceptualize our duties to nonhumans. Just as we want our human community to thrive, we also should want the plant and animal communities we live with to be healthy and flourishing. We may even believe we have a duty to help these plants and animals to flourish, or at least not to interfere with their lives. Leopold calls this moral sensibility a *land ethic.* The concept of a land ethic helps to explain why Judy K. might feel responsible for doing something about the purple loosestrife in her pond: She senses that she has some sort of duty not just to her neighbors but toward the land itself. We'll explore what exactly that duty is in later chapters. Our point here is simply that this sense of connection to the land, of being part of a biotic community, can be a valuable part of a good, choiceworthy life.

Thinking of the moral community as including nonhumans challenges the traditional assumption in ethics that only humans deserve moral consideration. But that challenge may not go deep enough: Notice that Leopold and Sylvan still focus on the local community and the immediate effects of one's actions—the geographic and temporal scales they're working with are the same as traditional ethicists use. They have not yet grappled with the fact that we are also involved in global systems, so our actions may affect people in very distant places, people with whom we don't feel any sense of community. And the pollution we put into that system might linger for generations, affecting humans and nonhumans in the distant future. This raises the possibility that environmental ethics for the twenty-first century may have to rethink the idea of moral community. Later chapters will explore what this new, twenty-first century moral community might look like.

2.6 Schools of Environmental Ethics

The landscape of environmental ethics is divided into *anthropocentric* and *ecocentric* systems. "Anthropocentric" means human-centered. These ethics are based on the premise that only human life has *intrinsic value* (that is, it is good in itself);

nonhuman life is valuable only to the extent it contributes to human goods.[1] Ecocentric ethics recognize that nonhuman life also can have intrinsic value; it can be an end in itself. The more radical ecocentrics argue that human life has no greater moral value than nonhuman life; all living things have equal value. Two influential schools of ecocentric ethics are Deep Ecology and Ecofeminism.

2.6.1 Deep Ecology

Founded by Swedish philosopher Arne Naess, this school conceptualizes humans as parts of a greater ecological system that has value in itself. Ethical reasoning begins when the individual develops an ecological consciousness, an awareness of being part of this larger planetary biosphere. Individuals who achieve this awareness will understand intuitively that if we harm nature we are harming ourselves. Deep ecology therefore rests on the principle of biocentric equality — all things in the biosphere have an equal right to live and reach their individual forms of self-realization. The practical implication is that we should seek to minimize our impact on other species and on Earth in general.

2.6.2 Ecofeminism

Ecofeminists identify relationships of domination — of men over women and humans over nature — as the chief source of environmental damage and injustice. They argue that in Western societies, the feminine and the natural are closely identified with each other, and both are conceptualized as naturally passive and subordinate to the goals and projects of men. Thus these two systems of domination are deeply interrelated; ecofeminism seeks to overthrow both of them.

Ecofeminism suggests that even if your primary concern is social justice rather than our relationship to nature, you may find that environmental ethics is important. For example, the African American social justice advocate Carl Anthony has argued that people subject to oppression can find meaning and solace in environmental ethics. He suggests that racial oppression can deprive its victims of a secure sense of self and alienate them from the larger community. They are made to feel as though they are "strangers in a strange land." Anthony found that studying environmental science helped to combat this destructive mindset by giving him a different perspective on his society — a perspective that does not simply reflect the values of White civilization. He writes, "The knowledge of the earth, and of our place in its long evolution, can give us a sense of identity and belonging that can act as a corrective to the hubris and pride that have been weapons of our oppressors" (Anthony 2006, p. 203.)

[1] The concepts of intrinsic and instrumental value are discussed further in Sect. 4.2.1.

2.7 Ethics and Social Change

Even if we design a better environmental ethic for the twenty-first century, some people working in environmental policy won't be interested in what ethicists have to say. The problem is this: Leopold, Sylvan, and many other environmental ethicists seem to assume that developing an environmental ethic will make us better stewards of our natural environment. Many social scientists are skeptical. It's natural to believe that our ethics shape our individual behavior, and therefore changing individual ethics will result in better behavior. But there are also good reasons to believe that transforming individual ethical beliefs is not the best way to make large-scale social changes. There are systemic forces that shape our behavior, and these may swamp individual ethics. In fact, many social scientists argue that individuals actually have very little moral agency; they are embedded in deeply unjust political and economic systems that deprive them of morally meaningful choices. This is an important disciplinary difference between traditional ethics and social sciences like sociology and political science: Ethicists usually focus on individual decisions, but social scientists focus on understanding the *social system*, because that is what shapes individual behavior.

Surely there's some truth to this systemic perspective. Consider the choices you have about how to use natural resources: Do you have any control over the source of the energy you use to wash your clothes or cook your food? How much control do you have over how your food is grown? We are often like Judy K., facing an invasion of purple loosestrife that she didn't cause and can't fix, at least not on her own. This is a systemic problem over which individuals acting alone have little control.

But even these social scientists usually don't conclude that individuals can't do *anything* useful at all. For example, understanding the defects in our social systems may lead us to political action to change those systems. Perhaps Judy needs to organize the neighborhood, to develop a consensus about which exotic species need to be controlled and develop a plan for doing so. In other words, if we're stuck in an unjust system, maybe the ethical thing to do is to change the system. We may need to develop a political ethic in addition to our environmental ethic.

Moreover, even if political action fails and our choices remain constrained by our social context, we can still try to bring our practices and policies into line with the best understanding of our environmental values. An environmental ethic may or may not improve the world, but it can certainly help you live a more meaningful personal and professional life in relationship to your biotic community.

Further Reading

Anthony, C.: Reflections on the purposes and meanings African American environmental history. In: Glave, D., Stoll, M. (eds.) To Love the Wind and the Rain: African Americans and Environmental History. University of Pittsburgh Press, Pittsburgh (2006)

Brennan, A., Lo, Y.S.: Understanding Environmental Philosophy. Acumen Publishing, Durham (2010)

Callicott, B.: Companion to a Sand County Almanac. University of Wisconsin Press, Madison (1987)

Jamieson, D.: Ethics and the Environment: An Introduction. Cambridge University Press, Cambridge (2008)

Kheel, M.: Nature Ethics: An Ecofeminist Perspective. Rowman & Littlefield, Lanham (2008)

Leopold, A., Sand County, A.: Almanac with Essays on Conservation from Round River. Ballantine Books, New York (1966)

Light, A., Katz, E. (eds.): Environmental Pragmatism. Routledge, New York (1996)

Pojman, L.: Ethics: Discovering Right and Wrong, 4th edn. Wadsworth, Belmont, CA (2002)

Murphy, J.: Santeria: An African Religion in America. Beacon Hill Press, Boston (1988)

Taylor, P.: Respect for Nature. Princeton University Press, Princeton (1986)

Chapter 3
Justice and Political Duties

Contents

3.1 The Problem of Pond Maintenance

There is a small pond on Spring Lakes that some of the tenants use for fishing and swimming. Maintaining the pond free of weeds and algae can be expensive, and Judy K. is considering covering that cost by charging the tenants a fee. But who should be charged? Only some of the tenants use the pond for swimming or fishing, but they all benefit from its aesthetic value. It also provides important ecosystem services — food and habitat for wildlife, for example — and Judy thinks we all have a duty to contribute to maintaining a healthy ecosystem. But some of the tenants have very limited incomes, and they argue that those who use the ponds more should pay more. Other tenants argue that Judy should pay the full cost of maintaining the ponds because she (as the owner) benefits the most from the added value that healthy ponds bring to the property. What is the fair approach?

K. K. Smith, *Exploring Environmental Ethics*,
AESS Interdisciplinary Environmental Studies and Sciences Series,
https://doi.org/10.1007/978-3-319-77395-7_3

3.2 Two Theories of Justice

The maintenance fee issue is a question about justice, which is where we will begin our exploration of environmental ethics. Why justice? Why not begin with compassion, or courage, or resourcefulness? There are many virtues that we might want to realize in our lives, but for most people justice is always one of the most important. We want to be fair, to treat others as they have a right to be treated. Justice is also critically important when we're thinking about how we are going to collectively manage our natural resources. Justice is a primary obligation of government and citizens. We don't always expect political actors to be kind or charitable, but we do expect them at least to be just. But what exactly do we mean by that?

The Greek philosopher Aristotle, in his treatise *The Politics*, tells us that justice is a sort of equality. But justice, he cautions, does not mean treating everyone the same. Rather, justice requires us to treat equals equally and unequals unequally. For example, if all the students in a math class were given a "C", they would be treated equally — but this wouldn't be just. Students who prove themselves to be better mathematicians should receive a higher grade, shouldn't they? Or maybe not: maybe students who work harder should receive a higher grade, even if they aren't as talented as some of the other students.

You see that although we may agree with Aristotle's definition of justice, we run into difficulty when we try to apply it. We have to ask ourselves, "Equal in what respect?" What characteristic or criterion are we using to award grades, privileges, and other goods and harms? Maybe this isn't too difficult to decide in the case of the math class: We know that the purpose of the math class is to teach people how to do math, so we should reward either hard work or talent or both. The purpose of the class helps us narrow down our options. But the question "Equal with respect to what?" is more difficult when we're trying to decide how to distribute general social goods, like political rights, wealth, clean air, or access to parks.

Following the math class example, perhaps we should ask, "What is the purpose of our society?" Maybe our purpose is to pursue wealth; in that case, we should reward wealth and efforts to acquire wealth. Wealthy people should be given lots of political power, which they would probably use to make sure that parks are built in their neighborhoods and hazardous waste facilities aren't. Poorer people would be motivated to work hard to become wealthy. And this would seem fair because our collective goal would be to create a very wealthy society.

But does our society really have a collective, overarching purpose, in that sense? What about Judy K.'s little community: Does that small society have a purpose, some ultimate goal that could help her decide how to distribute goods and harms? Maybe the purpose of society is simply to maintain the conditions in which people can pursue their own individual purposes and values. If so, how do we decide who should receive advantages and disadvantages?

Many political philosophers have addressed that question. English philosopher Jeremy Bentham (1748–1832), writing toward the end of the eighteenth century, offered a compelling and admirably simple answer: The purpose of human society is to increase happiness and reduce suffering. In his view, everyone is equal with respect to the importance of their happiness and suffering. So justice means treating everyone's suffering as equal, and trying to minimize it. (Incidentally, Bentham included animals' suffering, since animals can suffer as well — although they might suffer at different intensities and from different kinds of harms.) A whole school of philosophers, called *utilitarians*, have followed and expanded on Bentham's theory.

Utilitarianism is one of a family of ethical theories called "consequentialism." Consequentialists evaluate actions or rules based on their likely consequences — whatever consequences they consider morally relevant. Utilitarians believe that the consequence to aim for is increasing "utility", which might be defined as happiness, satisfaction of preferences, or objective welfare. Utilitarianism was developed specifically to help lawmakers evaluate public policy, and it remains the dominant ethical framework in policy arenas.

Utilitarians argue that a just social system would be designed to reduce human (and maybe animal) suffering, treating everyone's suffering as equally important. A utilitarian deciding where to put a park might say it should be built wherever it's cheapest and the most people can benefit from it. Or, the utilitarian might conclude that what makes the most people happy is to let everyone who would be affected participate in deciding where to put the park.

If Judy K. were a utilitarian, she would consider what approach to maintaining the pond would make the most people happy. But that could lead to an odd result: What if one tenant is very unpopular, and the rest of the tenants would be wildly happy if he were required to pay to maintain the pond? That might lead to the greatest amount of happiness, but it hardly seems fair.

"Rights" theorists would agree that it isn't fair. They argue that humans have certain fundamental rights that can't be abridged, even if it would increase the total amount of happiness in society. (This kind of argument is also commonly called *deontological,* from the Greek term for "duty"). Rights theorists are less concerned with happiness and more concerned with *dignity*. Respecting human dignity, they argue, requires protecting *rational autonomy*: humans' ability to govern their own lives, to make rational decisions about how to live. Under this view, rational autonomy is the essence of freedom and dignity. It is what makes human life morally valuable. Therefore, a just society must be designed to respect this autonomy. That is the purpose of rights: to create a sphere of personal liberty that cannot be abridged by others without a good reason. For example, a rights theorist might argue that everyone who is affected has a right to participate in deciding where to build a park, even if that procedure leads to a lot of conflict and results in no park being built at all! Respecting rights is more important than building parks. Rights theorists would also argue that everyone has a right to be treated equally, unless there's a good reason to treat someone differently. Of course, that leads us back to the question of what constitutes a good reason to treat someone differently, which depends on the purpose of society. For rights theorists, the purpose of society is not to maximize happiness but to respect human dignity by protecting rights (that is, personal liberty or autonomy).

One of the most common ways to develop this "rights" view of justice is through the idea that political society is a *social contract*. Social contract theorists invite us to imagine a group of people living in a "state of nature" — that is, in a state of anarchy, without any government to make and enforce laws. Everyone would have a great deal of freedom; their natural rights wouldn't be abridged by any laws. But they would soon discover that this state of perfect freedom is undesirable. Others would be free to infringe their rights, to dominate and terrorize them. So reasonable people would come together and voluntarily agree to form a government whose purpose is to protect our rights: This is the social contract. True, very few governments actually originated this way, but it's fair to say that if a government acts as though it's fulfilling a reasonable social contract, then it's a legitimate and just government and we should obey it. Your rights, then, are the rights that reasonable people would agree that government should protect.

Two Social Contracts

The term "social contract" is used widely inside and outside of academics; no one discipline "owns" the term. Moral philosophers sometimes use the term to explain our private, individual moral duties to any other moral actor. They reason that we must respect the rights of other persons because of an implicit "contract": you don't hurt me and I won't hurt you. This social contract is meant to be universal, encompassing any other moral actor (perhaps even nonhumans).

This use of social contract theory is falling out of favor with moral philosophers, because it seems a very narrow and limited explanation of why we have duties toward others. (Do you really refrain from hurting people only because you don't want them to hurt you?) But even if you reject social contract theory as an explanation for your obligations to other private individuals, it may still be useful as a way to explain political obligation — which is how we're using it here.

3.3 The Social Contract and the Limits of Government

Social contract theory is one of the most influential theories of political obligation in liberal states (that is, states that are based on the assumption that humans are inherently free and deserve equal rights).[1] A theory of political obligation is a theory that explains why citizens are obligated to obey the law. Social contract theory is not the only such theory, but it is a common and familiar one. So we are going to use social contract theory as a basic framework for thinking about justice and political duties in the realm of environmental policy.

[1] Liberal states include the United States, most of the European states, and many other states around the world that were influenced by the Western Enlightenment.

The social contract idea is useful for two purposes: It tells us what the government can legitimately do to citizens (the limits of government power) and it tells us what citizens should be willing to do for the political community (the extent of civic obligation).

Under social contract theory, protecting the citizens' rights is the *only* legitimate and reasonable excuse for the government to constrain our liberty. It's the only use of government power that reasonable people would voluntarily agree to. Because people have different moral, religious, and philosophical views, they wouldn't all agree to a government that attempts to make us more religious or moral or to achieve some specific vision of the good life. Social contract theorists assume that we would prefer government to just focus on protecting our basic rights and leave us free to pursue our own visions of the good life. In short, social contract theory is a theory of *limited government.*

Obviously, this theory poses a challenge if we expect the government to try to make us more environmentally responsible. Liberal theory suggests that government can try to shape citizens' values only to a limited extent. Basically, it can encourage citizens to respect one another's rights. But as soon as the government tries to make us "good" (more religious, more moral, or more environmentally sensitive), it is open to the challenge that it's exceeding its proper authority. Liberal governments are supposed to remain neutral among different conceptions of the good life. (Political theorist Cary Coglianese offers a good discussion of how this concept of liberal neutrality affects environmental policy in "Implication of Liberal Neutrality for Environmental Policy.")

Social contract theory also teaches us that citizens have only limited obligations to government. Citizens are expected to obey the law, but only as long as the law is reasonable and doesn't exceed the proper sphere of government power. If the government tries to invade our rights without good reason, we are allowed — even expected — to resist.

The fact that there are limits to what the government can ask of us is very important for environmental policy, as we will see below. But we can also use social contract theory to think about how a private citizen, like Judy K., should treat others in her community. Perhaps instead of thinking of herself as simply a businesswoman trying to make a profit, she wants her tenants to feel that they are part of a choiceworthy community — that is, a community in which they are respected and valued, a community bound by a social contract. That is why she cares about justice. Being fair to others is one of the basic ways we show respect for one another.

Social contract theory is useful for thinking through basic questions of justice. But notice that the idea of the social contract, like the idea of moral community, has a scale problem. How far, geographically and temporally, does the social contract extend? Should Judy K. include her neighbors in the social contract? What about past and future generations? Because environmental impacts often extend beyond the immediate social community (and may linger for centuries), environmentalists are tempted to expand the social contract to include anyone (including nonhumans) who may be affected by a policy decision. But expanding the social contract also expands the reach of government—which may conflict with the liberal ideal of limited government. We'll return that point below, when we discuss government's duty toward the environment. But first we will consider an important question that social contract theory raises: Do we have a right to a clean, healthy environment?

The idea of human rights is based on the view that all humans have an equal claim to dignity. Human rights are those rights that must be respected if one is to live a life of basic dignity. Many human rights tell us what the government should *not* do to people: It shouldn't punish them without due process and a fair trial; it shouldn't restrict their freedom of speech, or tell people which god to worship, or force them to marry, for example. These are sometimes called *negative rights* because they are aimed at restricting government action. The rise of the welfare state in the twentieth century has led some people to claim additional *positive rights*: rights to social goods that the government should provide directly or at least ensure that citizens have access to (through well-managed markets and a well-functioning economic system). These positive rights might include the right to food, housing, health care, education, and a job.

Even more recently, some environmental activists have claimed that there are environmental human rights. That is, they argue that the government should ensure the people have access to a healthy and safe environment, as well as access to natural resources (like water) that are necessary for human life. Environmental human rights, they argue, are implied by humans' basic right to life. After all, how can we live without a healthy environment?

Environmental human rights are positive rights, or rights to social goods that the government should provide. Critics of the concept of environmental rights argue that it's not a good idea to cast claims to government resources as basic human rights. While a negative right (like freedom from arbitrary arrest) is something you either have or you don't, clean water is something you can always have more of. How much and how clean must the water be? And how do we balance demands for a clean environment against all the other goods that we want government to provide? Calling clean water or clean air a "right" can be seen as an illegitimate move in a policy debate — an attempt to tip the balance in favor of your preferred good over other important goods like education or health care.

But lawsuits in international forums have had some success in winning recognition of environmental human rights. These lawsuits have helped to force some governments to put more resources into environmental protection. So the concept of environmental human rights seems to have political value, even if they are philosophically questionable.

3.4 Distributive, Procedural, and Structural Justice

Achieving a just community can be quite complicated. Judy K. wants to make sure that the pond maintenance fee is distributed fairly among her tenants. That is, she wants to achieve *distributive justice*. This is one way to treat everyone as equally deserving of respect. But she might also decide that instead of simply imposing a fee on the tenants, she should create a process for her tenants to participate in to decide about the fee. Allowing people to participate in making the rules they will live under is another way to show equal respect for everyone. Coming up with a fair procedure for distributing goods and harms is called *procedural justice*. The clever reader will immediately see the problem: What if the tenants themselves aren't very fair-minded? If they are allowed to decide how to distribute the fee, they may decide to impose it on the most unpopular tenant. The decision process might be fair, but the outcome isn't.

There are ways to design processes that reduce the likelihood of unfair distributive outcomes. For example, Judy might require a unanimous decision, so no single individual can be unfairly targeted. Sometimes it's a good idea to build into the process a lot of discussion, so everyone's concerns and perspectives can be taken into account. This deliberation is often useful in making people feel that they are part of a community, so they may be more accommodating and fair in their decision making. That is, procedural justice is important not only because it treats everyone as equally deserving of respect; it also plays an important role in educating people into the value of justice. It can make us better citizens.

Still, in a small community, it's hard to ensure that personal likes and dislikes won't lead to unfair outcomes. In a large community, personal conflicts may not have as much effect on decisions. But we often find other sources of unfairness: racial, gender, and other forms of prejudice can lead to *structural injustice*. Understanding structural injustice is critical to understanding the challenges of creating policies (including environmental policies) that are fair to everyone.

The idea of structural injustice is not too difficult to understand: It simply means that social systems and practices favor some groups over others, so that policies that apparently treat everyone the same in fact favor some groups over others.

Imagine, for example, a society in which ordinary citizens fought hard to win the right to own firearms. Imagine that this fight resulted in laws that explicitly grant every adult male the right to hunt. After winning this right, the men in this society hunted for the family while women tended gardens. Over time, the men developed traditions — formal societies and informal practices — to educate their sons into the art of hunting. Going on one's first hunt became an important rite of passage into manhood. Now consider what might happen in this society as the role of women changes: As women enter into fuller social equality, they too might seek the right to hunt. Perhaps they will succeed in changing the law to ensure that every adult male and female has the right to hunt. In other words, they will finally have legal equality. Would this legal equality mean that every woman has an equal chance to become a hunter?

There's reason to believe that it wouldn't. Even though women have an equal legal right to hunt, they grow up in a society that is organized to initiate men, but not women, into hunting. Women may still learn early on that it isn't feminine to hunt, that hunting is for boys. Men may not be interested in taking girls hunting, or they might simply assume that girls wouldn't be interested in hunting. Hunting societies may not want to change their practices to let in women. The laws may be equal, but the society is not: It is structured in such a way that inequalities will persist. This is the situation we call "structural injustice."

Structural injustice is present when past injustices continue to shape our social, economic, and political institutions and practices. Issues of structural injustice have been the special focus of the *environmental justice movement*, which has argued that structural injustices lead to the inequitable distribution of environmental harms (like pollution) and environmental goods (like parks). The environmental justice movement began in the 1980s in the United States, drawing attention to evidence that racial minorities are disproportionately affected by hazardous waste sites and other undesirable land uses. It has grown into an international movement that critiques Western colonialism, racism, and patriarchy on environmental grounds.

In the United States, environmental justice advocates have been concerned about several kinds of structural injustices: patriarchy (leading to gender inequality), white supremacy (leading to racial inequality), capitalism (leading to class inequality), and industrialism (leading to the domination of nature by humans). One influential social critic, Murray Bookchin, argues that a basic injustice in all modern societies is the very concept of social hierarchy: "Society is poisoned. It has been poisoned for thousands of years, from before the Bronze Age. It has been warped by rule by elders, by patriarchy, by warriors, by hierarchies of all sorts." He argues that we must get rid of our practices of domination and replace them with an "ethics of complementarity" that "respects other forms of life for their own sake and that responds actively in the form of a creative, loving, and supportive symbiosis" (Bookchin and Foreman 1991, p. 33).

From this perspective, achieving just outcomes in a society with many structural injustices may require more than simply ensuring fair procedures; it may require changing basic social practices and providing greater support to those who are systematically disadvantaged. It may also involve subjecting our basic beliefs and values to critical analysis, uncovering and correcting unconscious prejudices. It may even require radical social and ethical transformation. This is clear when we're dealing with issues of racial or gender justice, but it may be true for environmental issues as well. Our ideas about and relationships with the natural world may be warped by our desire to dominate others and to dominate nature itself.

Note that the idea of structural injustice requires us to move from the individual focus of traditional ethics to the systemic perspective we see in the social sciences. When we think about structural injustice, we have to abandon some core ideas of moral philosophy. Most importantly, structural injustice might be present *even if no individual actor is behaving unfairly*. No one is violating anyone's legal rights, and the victims may not even be aware that the system is working to disadvantage them. The concept of structural injustice therefore complicates our ideas about moral responsibility: If the injustice lies in the system, not in individual behavior, then maybe no one is at fault — and yet we may all share responsibility for changing the system. (This point relates to the idea of collective responsibility, discussed in Sect. 3.7.1 below).

3.5 Pond Maintenance Revisited

What does all this suggest for Judy K's maintenance fee dilemma? At the very least, it suggests some things for her to consider. First, she may want to set up a decision making procedure that would allow the tenants to deliberate on the issue. But she may also want to set some guidelines for that deliberation to make sure that the outcome (and not just the process) is fair. She should probably consider whether structural injustices in the larger society may play into the deliberation. One of her tenants is a black man who has had difficulty finding work in the predominately white community surrounding Spring Lakes. One is an unmarried woman with a child. They are both facing severe financial challenges. Will the other tenants treat them fairly? Some of the tenants are very reserved and clearly uncomfortable making arguments in front of a crowd. Others have very strong views about what would be fair. What sort of decision making process would make everyone feel respected and lead to a just outcome? What considerations should the tenants take into account in making their decision? All of these questions are important to designing a fair decision making process.

3.6 Duties of Government

So far we've been thinking about what duties Judy K. may have towards other humans in her community. But what about her duties toward nonhumans? Recall from the last chapter Aldo Leopold's concept of the biotic community: Judy might consider that she is part of a biotic community that includes all the nonhuman entities that are part of a healthy ecosystem. She may think that has a duty to attend to their welfare as well. Extending justice to nonhumans raises some interesting questions which we will address in the next chapter. For now, let's think about what the biotic community idea means for social contract theory and the government's duty of justice.

Perhaps you've decided that we are members of a biotic community, and that humans do have duties to animals and even ecosystems. Does it follow that the government ought to enforce these duties? Should the government punish those who abuse animals? Should it ban the sale of meat for human consumption, or prevent the destruction of wetlands?

There may be many good reasons to enact such policies. Perhaps they are justified by the government's concern for human welfare and happiness. We can justify a wide range of environmental regulations simply on the grounds that such regulations serve the interests of the human community. A clean environment helps its human inhabitants remain healthy and happy. But can we justify restricting human freedom solely to protect nonhuman welfare for its own sake? In other words, should the government look out for the welfare of nonhuman entities, even when it conflicts with human welfare?

There's a strong argument that a government based on liberal principles should *not* try to protect nonhuman entities for their own sake. While we may have private moral duties to those entities, it may not be a proper use of government power to enforce those duties. Remember that the basic idea of the social contract is that people agree to obey the government on the condition that the government protects their liberty, and doesn't try to impose other people's religious or moral views on us. Protecting nonhumans might look like one of those moral views: It's a perfectly sound moral principle, just like the principle that we should normally try to tell the truth. But that doesn't mean we want the government to enforce it. We expect the government to leave us free to make our own moral decisions, unless we're violating someone's rights. That, after all, is the purpose of a liberal government: to protect human rights, not to protect plants and animals.

This line of argument is why some people argue that nonhumans have rights, too. It's much easier to justify using government power to protect animals (for example) if we believe animals have rights. However, even if animals do have rights, it doesn't really follow that our government should protect those rights. No government sets out to protect *everyone's* rights, after all. For example, it's not the job of the United States government to protect the rights of Japanese citizens living in Japan. That's the job of the Japanese government. We might reason, along the same lines, that it's not the job of the United States government to protect the rights of wolves. Wolves, perhaps, are best considered members of some other community, not the human community that the United States government is supposed to protect.

For Aldo Leopold, though, wolves *are* members of our community—our biotic community. His argument for extending the moral community to include nonhumans might also be used to extend the political community, and the reach of government, to protect those beings. The idea of a social contract that includes the biotic community is appealing. It captures the intuition that nonhumans deserve respect and consideration in their own right; they don't exist merely for our purposes. But (as discussed above) the idea that all nonhuman animals (and plants as well?) have rights also risks extending government power much farther than liberal citizens are accustomed to. Should there be laws governing when we can swat a mosquito or kill a mouse?

Of course, many countries have laws that protect animal welfare. In the United States, these laws apply primarily to domestic animals: pets, farm animals, and animals we keep for entertainment or research. We might conclude that these animals are essentially members of our society: We have taken responsibility for their welfare, so we have a collective duty—as citizens—to make sure they are well-treated. In contrast, our duties to wild animals are private ones: They may be critically important to our ethical lives, but perhaps we don't expect or want the government to enforce them.

The key point is this: Expanding the scale of government power to match the scale of environmental problems sounds reasonable, but it carries the risk of creating a much bigger and more powerful government than we really want. There are and should be limits to legitimate government power in a liberal regime. If we're going to use laws to restrict a person's liberty with respect to the nonhuman world, we need a strong justification. We need to be clear about which nonhumans are members of our political community—our social contract—and therefore deserve government protection for their own sakes.

Constitutional Rights for Nature

In 2002, Germany amended its constitution to specify that the government has responsibility for protecting the natural foundations of life and animals in the interest of future generations. Is it fair to conclude that the German people have agreed to a social contract that allows the government to restrict human liberty in order to protect animal welfare?

Not necessarily. First, it's not entirely clear whether this provision means that the government can protect animals for their own sake or only to the extent they serve human interests. Second, the amendment was adopted by the German legislature, not by a popular referendum, and it wasn't unanimous. So we can't say that all the citizens actually agreed to it. Of course, we might reason that German citizens agreed to be bound by the decisions of the majority. But even so, the majority should still assure themselves that imposing obligations on people who didn't consent to them is reasonable — that they *should have* consented, or at least that they had good reasons to consent. So the amendment probably won't end debate over the extent of government authority to protect nonhumans.

Ecuador and Switzerland are also among the countries that have written something like "rights for nature" into their constitutions. The activist group Pachamama Alliance argues that nations should "recognize the Earth and all its ecosystems as a living being with inalienable rights: to exist, to live free of cruel treatment, to maintain vital processes necessary for the harmonious balance that supports all life" (Pachamama Alliance, http://www.pachamama.org/advocacy/rights-of-nature (2018), accessed Jan 26, 2018).

This international constitutional movement probably reflects the fact that most constitutions were written before the age of environmentalism. Therefore, they don't address the government's authority with regard to the environment. The United States Constitution, for example, does not explicitly authorize the federal government to protect the environment. Most federal environmental regulations were enacted under Congress' powers to regulate interstate commerce and manage federal property. State governments have authority to protect the environment under their general power to protect the health and welfare of their citizens. But this power, too, is limited by Americans' traditional strong protection for property rights.

Should the United States Constitution be amended to recognize rights of nature? It's not clear what legal or policy impacts such an amendment would have. But perhaps a national conversation could help to clarify how far we are willing to extend our social contract.

3.7 Duties of Citizens

Under traditional social contract theory, once we've decided what the government can legitimately require citizens to do, we also know what citizens are obligated to do: They are obligated to obey the law. But many political theorists point out that this limited view of civic obligation isn't really workable. We actually require more of citizens than simply obeying the law. For example, we expect citizens to contribute to the common good when they can — to be engaged in civic life, to participate in politics (to some extent), and to raise their children to be good citizens. We expect people to be tolerant and respectful of one another, beyond the minimum required by law. In fact, in a society that is troubled by structural inequality, it may be necessary for citizens to engage in reparative work, to restore relations of trust and harmony between the oppressed and privileged.

3.7.1 *Restorative Justice*

That duty to make society more just is often called *restorative justice*, which we can add to our growing list of different kinds of justice. Theories of restorative justice were developed to help us navigate societies troubled by historical injustice, where social groups don't trust one another. Restorative justice aims at restoring relationships of trust between the oppressed and the oppressors. It usually involves those who have perpetrated injustice (or benefitted from injustice) taking responsibility for their part in creating these troubled relationships. The victims also have a duty to forgive and to work to repair the community. South Africa's famous Truth and Reconciliation Commission is an example of restorative justice in practice. The commission worked to repair race relationships after the end of apartheid. Victims of human rights violations were invited to give statements about their experiences, and some were selected for public hearings. Perpetrators of violence also gave testimony and received amnesty from prosecution.

Engaging in restorative justice isn't the only thing we may expect of our fellow citizens. We also expect them to help each other out when disasters strike, and to manage natural resources wisely for future generations. For example, in August of 2005, a hurricane landed near the city of New Orleans and the levees that were supposed to protect the city failed. The resulting flood and inadequate government response resulted in thousands of deaths and the displacement of hundreds of thousands of American citizens. The disaster starkly revealed deep race and class inequalities in the United States that many Americans find very troubling. But most of us American citizens (including myself) experienced this event at a distance. We watched the images on television and wondered, what should *I* do about this? As an American citizen, what responsibility do I have for the failed levees, the poor rescue effort, the racial and class inequalities that made the whole disaster so much worse?

The idea of citizenship seems to imply the idea of collective responsibility. That is, there are some responsibilities you have not because of your individual actions but by virtue of being a member of a community. Imagine, for example, that you are one of hundreds of good swimmers on a beach without a lifeguard, and someone in the water is drowning. Do you have a duty to rescue the drowning person? If you don't, then no one else does either. But if you all try to rescue the person, you may get in each others' way. It's more sensible to conclude that you all have a duty to *organize* a rescue effort. That's what government does: It helps us fulfill our obligations to help one another.

But the government doesn't try to help everyone in the world, just our fellow citizens. Why should our fellow citizens get this attention, when they may be total strangers to us? One response is the social contract: The government is the result of our implicit social contract with one another. But the idea of the social contract doesn't really explain why we feel obligated to fellow citizens. After all, I never *actually* negotiated a contract with the citizens of New Orleans. Why are they members of the social contract, when people in Canada (who happen to live much closer to me) are not?

The philosopher Robert Goodin has one answer to this question. He argues that members of the same political society have special obligations to one another because we are more *vulnerable* to one another. Members of the same group or organization can do more harm to one another than mere isolated strangers can. The residents of New Orleans had come to depend on the United States government — on their fellow citizens — for aid. They assumed they would receive the same sort of disaster assistance that other communities in the United States receive. Therefore, the rest of the country had a special responsibility to provide that aid.

3.7.2 Ecological Citizenship

Goodin's argument explains why we feel obligated to help our fellow citizens: Our government is like a rescue organization, and we are all responsible for seeing that it operates properly. But his logic can extend further: philosopher Andrew Dobson argues that we have special responsibilities to *anyone* who is particularly vulnerable to our actions. He would therefore extend our duties of citizenship beyond the borders of the nation, to citizens and even nonhumans who suffer as the result of our government's decisions. Dobson urges us to adopt a concept of *ecological citizenship*, under which our ecological impacts on the world give rise to political obligations — obligations that we can fulfill only by living more sustainably.

Ecological citizenship is consistent with the idea of the biotic community and extending rights to nonhumans. And like those ideas, it threatens the idea of limited government: If our political obligations extend to any human or nonhuman who might suffer from our actions (even well into the future), is there any real limit on how far the government can go to protect all those beings? The idea of ecological citizenship suggests that our political obligations are as extensive as our private moral obligations. Maybe that's going too far. Traditionally, the duties of citizenship are more limited than the duties of a good person. The good citizen sees only the suffering of her fellow citizens; the good person sees the suffering of the whole planet.

3.7.3 Citizenship and Global Environmental Problems

Global environmental problems like climate change and biodiversity loss are particularly challenging for conventional ethical reasoning. As discussed in earlier chapters, our ethical theories were developed to help people decide small-scale questions in face-to-face communities, in situations where individual actions had clear, direct consequences for people you knew. Under these conditions, it's not hard to follow the general moral rule of not harming others. But the harms of climate change and biodiversity loss are different; they result from complex systems of industrial production, transportation, and consumption in which we all participate but no one in particular controls. They are wicked collective action problems: They're caused by billions of past and current people interacting with natural systems in ways that seem completely blameless and even praiseworthy. Moreover, changing your individual behavior won't solve them unless billions of others do as well (in which case, you probably wouldn't have to!) Unfortunately, many global problems have similar characteristics: they are caused by a critical mass of ordinary acts, the chain of causation is complex, and harms are diffuse and remote in both time and space. It seems that virtually everything we do is causing harm to others. What duties does a citizen have with respect to these problems?

Again, our problem lies in disciplinary assumptions: Ethics focuses on individual actors; it assumes that they can make choices that matter. But these global problems are systemic problems, and they require a systemic solution. That is, solving these problems requires us to change our political and economic systems, and that requires political action. The concepts of collective responsibility and ecological citizenship (which were developed primarily by political theorists) can help to motivate such collective action in response to climate change or biodiversity loss (not to mention other global problems, like poverty, malnutrition, and oppression).

But in the meantime, do you have a duty to reduce your individual contribution to climate change by changing consumption habits? To reduce your contribution dramatically would involve living an ascetic life, forgoing most of the pleasures of our consumer society. *Asceticism* has a long and respectable history, but most of its advocates choose asceticism not simply to avoid harming others but as a way to realize goods like spiritual enlightenment, a more authentic life, and a richer connection to nature and the community. To choose an ascetic life merely to avoid doing harm seems like an impoverished conception of the good life. Surely a good human life achieves something more than merely doing as little harm as possible! And as important as environmental responsibility is, it is probably not the only important value in your life. You may also value a rich family life, service to the community, artistic or athletic excellence, or the pursuit of scientific knowledge. Realizing other values will probably require you to participate in the economic system that is causing climate change and loss of biodiversity.

Still, even if we cannot escape these environmentally problematic systems altogether and still live a choiceworthy life, we can reduce our dependence on those systems and encourage others to do the same. The choice to do so expresses your values and embodies them in your life; it makes them meaningful. In short, living lightly on the earth is a way of putting your money, and your life, where your mouth is.

3.8 Duties of Corporations

Citizens are not only members of states; they are also members of all sorts of intermediate associations: families, churches, clubs, professional associations, and corporations. They may have duties as members of these organizations, over and above their duties as citizens. How does membership in a civic organization affect one's political duties?

There are a lot of topics we could discuss here, but I want to focus on one question that is particularly important in environmental affairs: What are the duties of corporations (both for-profit and non-profit) with respect to environmental sustainability? Would you as CEO of a major corporation have the same set of environmental duties that you would have as a private individual or a citizen? Or do your duties to the organization limit or change those duties?

This question is part of the debate about corporate social responsibility. An important entry in this debate was offered by American economist Milton Friedman in 1970. He argued forcefully that the only social responsibility of a corporation (beyond obeying the law) is to increase its profits. He reasoned that managers of these organizations have made legally enforceable promises to shareholders; the managers received investments on the understanding that they would carry out the corporate mission and try to make a profit for investors. It would be unethical to take that money to pursue some other end, like cleaning up the environment. Friedman acknowledged that a lot of things that corporations do for the common good are also good for the bottom line. But in his view, the corporate manager has no authority to pursue social projects that don't help to make profits.

R. Edward Freeman took up the debate in 1984, in his book *Strategic Management: A Stakeholder Theory*. Freeman disagreed with Friedman. He argued that corporations are created by the law to serve social purposes, and we can define their social obligations more broadly than mere profit-making if we choose. Freeman pointed out that corporations wield a great deal of social power; their decisions can affect millions of workers, consumers, and the general public. They should accept the additional responsibility that comes with such power. He therefore recommended that corporate managers consult all the various *stakeholders* in the firm when they make important decisions: shareholders, workers, customers, and anyone else who might be benefitted or harmed by the corporation. The managers should try as far as possible to serve all these interests. Environmentalists might add nonhuman stakeholders to the list.

There's no consensus in the business world on how far corporate social responsibility extends. Some still endorse Friedman's position that the corporation exists to serve its shareholders. But the debate took on new life in 2004, when

Paul Buchheit and Amit Patel, the founders of Google, issued their initial public offering. They let potential shareholders know that the company might forego short-term gains in order to serve social ideals, such as ensuring consumer privacy. But Buchheit and Patel, like many contemporary advocates of corporate social responsibility, insist that they are not acting against shareholder interests. Their IPO states, "We believe strongly that in the long term, we will be better served — as shareholders and in all other ways — by a company that does good things for the world even if we forgo some short term gains" (Google 2004, p. 27).

The debate over corporate social responsibility is also relevant to nonprofit corporations, like most colleges. These corporations also receive donations on the understanding that they will use the money to pursue the basic mission of the organization. College administrators, for example, might be reluctant to spend money on environmental sustainability unless it has some obvious connection to the central mission of the college: education. In sum, when you are making decisions on behalf of an organization, you may not be able to follow your private environmental ideals. Having promised to serve the ends of the organization, you may be constrained (legally and ethically) from using the organization's resources for other ends, even if you think those other ends are more important.

One lesson to be drawn from this discussion of corporate social responsibility is that you should be very careful about which organizations you join. When you join a corporation (by taking a job, for example), you may become ethically obligated to further its mission. So you should make sure its mission is consistent with your values.

3.9 Spring Lakes, Inc.

Earlier in this chapter, we considered whether Judy K. should set up a decision making procedure so that the tenants could determine a fair way to distribute the pond maintenance fee. But what if Judy K., along with some investors, decided to incorporate her rental business? Would letting the tenants participate in decision making still seem like a good idea? How would she explain that decision to her investors? Would she be justified in sacrificing some profits to ensure that the ecosystem remains healthy and her tenants are treated fairly?

It may be that Judy's decision about how to handle the maintenance fee will depend on how she understands her role: Is she a private property owner, a citizen, the governor of the Spring Lakes community, or a corporate officer? While the first part of this chapter discussed justice as though it were a general duty that applied the same way to everyone, the second half of the chapter suggests that the duty of justice might be shaped in large part by your organizational role or position. An officer of the state may have a strong duty to be just — but only up to the proper limits of government. As a citizen, you also have to respect those limits, even though, as a private individual, you may see beyond them. Finally, your duty of justice is affected by your roles in various organizations; your responsibilities to these organizations also shape your ethical world and your relationship to nature.

Further Reading

Aristotle: The Politics, 2nd ed. Trans. Carnes Lord. University of Chicago Press, Chicago (2013)
Bookchin, M., Foreman, D.: Defending the Earth. South End Press, Cambridge (1991)
Callahan, D.: A moral good, not a moral obsession. Hast. Cent. Rep. **14**(5), 40–42 (1984)
Coglianese, C.: Implication of liberal neutrality for environmental policy. Environ. Ethics. **20**(1), 41–59 (1998)
Davradou, M., Wood, P.: The promotion of individual autonomy in environmental ethics. Environ. Ethics. **22**(1), 73–84 (2000)
Dobson, A.: Citizenship and the Environment. Oxford University Press, Oxford (2003)
Fahlquist, J.: Moral responsibility for environmental problems—individual or institutional? J. Agr. Environ. Ethics. **22**, 109–124 (2009)
Freeman, E.: A Stakeholder Theory of the Modern Corporation. Cambridge Univ. Press, Cambridge (1984)
Friedman, M.: The social responsibility of business is to increase its profits. The New York Times Magazine, 13 September 1970.
Goodin, R.: Protecting the Vulnerable. University of Chicago Press, Chicago (1985)
Google, Inc: Final prospectus (filed 18 August 2004)
Picolotti, R., Taillant, J.D. (eds.): Linking Human Rights and the Environment. University of Arizona Press, Tucson (2003)
Sagoff, M.: Can environmentalists be liberals. Environ. Law. **16**, 775–796 (1986)
Walker, M.U.: Restorative justice and reparations. J. Soc. Philos. **37**(3), 377–395 (2006)

Chapter 4
Do We Have Duties to Nonhumans?

Contents

4.1 The Deer Problem

Judy K. has extensive gardens around her house. She also has a large herd of uninvited deer. Gardens and deer aren't very compatible. Despite Judy's wide array of deer repellents, they wreak havoc on her shrubs, flowers, and vegetables. But it's not just her gardens; the forest is also suffering damage from the deer.

It's not surprising that there are so many deer living here. The surrounding countryside is perfect habitat for deer, and most of the other property owners allow deer hunting. Judy has never liked the idea of having hunters wandering over her property, especially when small children were living there. But because she has never allowed hunting, the property has become a haven for deer. Her nephew and some of his friends like to bow-hunt; he has offered to keep the deer population down. Judy's neighbor Maria approves of the plan to allow deer hunting, which she thinks is necessary to protect the ecosystem (since humans have eliminated the deer's natural predators). But Judy's daughter thinks that bow-hunting is crueler than hunting with rifles, because bullets kill quicker and cleaner than arrows do. Judy worries that deer rifles would be more dangerous to the people living on the property than bow-hunting would be. She's sure that she has a duty to her tenants to make sure the property is safe. But does she have a duty to the deer? To the forest? To the other animals living in the forest? And how should she reconcile all these apparently conflicting duties?

K. K. Smith, *Exploring Environmental Ethics*,
AESS Interdisciplinary Environmental Studies and Sciences Series,
https://doi.org/10.1007/978-3-319-77395-7_4

4.2 Defining the Moral Community

Judy K. is struggling with a basic question in environmental ethics: To whom do we have moral duties? Put another way, who is a member of the *moral community*? Who (or what) has *moral status*? Traditional moral philosophy usually assumes that we have duties to other humans (or perhaps to all rational beings). The reason for this assumption varies; some argue that we respect other humans because we want them to respect us. This is an application of the social contract idea discussed in the previous chapter: We think of mutual respect as the result of an implicit bargain between two moral actors. Others focus on the fact that humans have something we value, like rational autonomy (the ability to govern our own lives rationally). We value that ability in ourselves, so we ought to value it in others.

Some philosophers, however, point out that we respect people who can't respect us in return and who don't have rational autonomy, like small children or people incapacitated by age or mental defects. Perhaps we respect these other humans simply because we recognize that their lives have value to them, just as our lives have value to us. But can't we say that the life of an animal also has value to it? Should we give animals the same respect we give to humans? And what about plants? Does a plant's life have value in itself, or only to the extent it's useful to rational beings?

4.2.1 *Value Theory*

We are now entering into an important part of ethics, value theory. Value theory concerns different types of values, what things or activities are valuable, whether value is objective or subjective, and related matters. We are not going to survey the whole field of value theory, but we do need to understand two concepts that appear frequently in environmental ethics: *intrinsic value* and *instrumental value*. Something has instrumental value if you value it simply because it helps you get something else. Money is a good example; you wouldn't value money if it weren't instrumental to getting other goods. In contrast, something has intrinsic value if it is good in itself. Happiness, joy, autonomy, and pleasure are all examples of things that some people consider intrinsically valuable. (Of course, some things have both instrumental and intrinsic value: an excellent gymnastics performance is good in itself, but it can also help you win a gold medal). Most people would argue that human life is intrinsically valuable; that is, we want to live just for the sake of living. But is nonhuman life intrinsically valuable? That is, does nonhuman life have value in itself, or is it simply instrumental to our (human) happiness and welfare?

Clever readers will notice that this is the question Richard Sylvan was raising in the last man scenario in Sect. 2.5: If you were the last person on earth, would it be morally acceptable to destroy all the other life on earth? If that life has only instrumental value — if it is valuable only because it serves to make human life better — then it might be acceptable. But if that life has intrinsic value — if it is valuable for its own sake — then you shouldn't wantonly destroy it. The issue is important in environmental ethics, because if nonhuman life has intrinsic value, then we should take steps to protect it, even (sometimes) at the expense of human welfare. This doesn't mean that nonhuman life would always, or even usually, outweigh the value of human life. But it does mean that Judy K. would have to take the value of the deer's life to itself into account when deciding how to protect her gardens and forest.

Does nonhuman life have intrinsic value? One way to approach that question is to think about the qualities that make human life valuable, and determine whether nonhuman lives have those qualities. That is a popular approach in environmental ethics, and we'll spend some time considering various arguments along these lines. But I will also introduce a different approach to the question offered by the philosopher Elizabeth Anderson. Her theory focuses more on *how* we value rather than *what* we value. In the next chapter, we will also take up a related question of great importance in environmental ethics: How should we value the lives of humans who aren't born yet? These two chapters should give you some useful ways to think about who belongs to your moral community.

4.2.2 Living Humans

According to Aldo Leopold, the time has come for us to extend our moral community to include the natural environment. He seems to assume that we're ready for this next step in the extension of ethics: We now recognize that all humans have equal moral status, so it's time to turn to nonhumans. But do we recognize all humans as moral equals? If not, does Leopold's advice divert our attention away from the importance of securing respect for all humans?

This is one of the common worries about environmental ethics, that it distracts us from addressing injustice toward humans. Of course, in some cases it might. If you have a limited amount of money to contribute to a charitable cause, you may have to choose between helping humans or helping nonhumans. But people concerned about the environment argue that we shouldn't be too concerned that prioritizing nonhumans would lead us to neglect our duties to humans.

First, there's little evidence that the rise of the animal rights and environmental movements has corresponded to a decline in attention to other social justice issues. On the contrary, concern with one sort of social problem often leads people to get involved with addressing other social problems.

Second, a number of theorists argue that failure to respect nonhuman life and failure to respect other humans are linked problems. They suggest that both failures stem from the same moral failing, a failure of ethical reasoning, empathy, or moral character. As we learn to care about nonhuman "others", we may become better at understanding and caring for human "others."

Finally, recall Murray Bookchin's argument (Sect. 3.4) that all social injustices stem from the same basic problem: Our addiction to social hierarchy. Bookchin argues that racial, class, gender, and other social hierarchies are linked to, and reinforce, human domination of nature. Under this view, addressing social justice requires us to adjust our practices toward and attitudes about the natural world as well.

For example, Alice Walker, an American writer and advocate for racial justice, sees no conflict between environmentalism and social justice. She begins with the premise that humans are connected to places and to nonhumans in intimate, vital relationships. One of the chief harms of racial injustice, for her, is that it destroys these connections. Injustice works to separate people from each other, but it can even disrupt bonds between humans and the land. As she explains in her book *In Search of Our Mother's Gardens*, "the history of all black Southerners is a history of dispossession. We loved the land and worked the land, but we never owned it" (Walker 1983, pp. 143–145). Black Southerners, she tells us, were expected to leave the South if they could. The brutality and violence of a racially unjust social order undermined their connection to the land. A major achievement of the civil rights movement was to convince black Southerners like Alice Walker that they had claim to the land of their birth. Under this view, achieving social justice is critical to helping a community create healthy connections to their natural environment.

4.2.3 Nonhuman Animals

Can we extend our sense of justice to nonhumans? It's usually not too difficult to convince people that they have some sort of moral duties toward *sentient* animals (that is, animals capable of feeling pain). Most people feel some obligation to avoid inflicting unnecessary suffering on animals. But we may not feel the same sense of obligation to wild animals or farm animals that we feel toward our pets. We may treat fish and reptiles differently than we treat mammals. Do these distinctions make sense? What do we owe nonhuman animals, and why?

Let's begin with the hardest case: Do animals have rights? Recall that rights are connected to the concept of rational autonomy. The purpose of rights is to protect our liberty to govern our own lives. Most animals probably don't have that kind of autonomy. They don't make life plans; they don't try to live a rational, meaningful life as humans do. That, at any rate, is the argument of philosopher R.G. Frey's well-known article, "Rights, Interests, Desires, and Beliefs." Frey argues that animals can't have rights because they don't have interests. Having interests means having desires — a desire to eat a sandwich, for example. But having that kind of desire requires you to have beliefs about the world (e.g., the belief that eating the sandwich would cause you pleasure.) And to have beliefs about the world, you have to be able to formulate ideas like, "That sandwich would taste good." Animals, he argues, lack linguistic ability, so they can't form beliefs. No beliefs means no desires, which means no interests, and therefore no rights. A good deal of work in animal rights is aimed at challenging Frey's reasoning.

Philosopher Tom Regan, for example, argues that nonhuman animals can have rights. Although animals may not be as fully autonomous as humans, they do have a *welfare*; things can go well or badly for them. More specifically, some animals (Regan focuses on mammals and some other sentient animals) have a sense of identity over time, they perceive and remember. Contrary to Frey's reasoning, they *do* have preferences, desires, and goals. They are "subjects of a life." We should respect their freedom to pursue those goals.

Regan's argument rests on the claim that animals can perceive, remember, have goals, etc. That claim has to be supported empirically. This is the work of studies in animal cognition and behavior, which reveal a wealth of diversity in animal minds. Some animals are highly intelligent problem-solvers, with self-awareness and the ability to interpret the mental states of other creatures. Some are even capable of deception, shame, and revenge. Of course, this research is always open to challenge. After all, it's difficult to know what's going on the mind of a creature radically different from ourselves. For that matter, it can be difficult to know what's going on in the mind of another human, or even our own minds! But researchers in this field argue that such difficulties don't prevent us from making reasonable inferences from an animal's behavior to its mental states and mental abilities, just as we do for humans.

Even if you don't agree that animals have the kind of rational capacity one needs in order to have rights, you might agree with utilitarians like philosopher Peter Singer that the ability to suffer can be the basis of moral status. Singer argues that we have a duty to reduce animal suffering just as we have a duty to reduce human suffering. He agrees, though, that an animal's rational capacity is relevant to how we weigh animal against human suffering. Because humans are cognitively and emotionally complex, their capacity to suffer is greater than that of most other animals. Humans, for example, can imagine the future and become invested in it. That's why, for Singer, the death of a human is a greater tragedy than the death of a cat. The human has lost a future in which she was deeply invested. Still, the death of a cat matters. For Singer, cats have the same moral status as humans.

4.2.4 All Living Things

Regan and Singer did important work in developing a philosophical case for the moral status of sentient animals. But sentient animals are still just a small fraction of all the nonhuman beings that environmental ethics is concerned with. These range from viruses and bacteria to plants and animals. These are all very different sorts of entities, of course, but they do have one thing in common: They are alive. Some environmental ethicists argue that this is enough to warrant our respect. For example, Paul Taylor argued in his book *Respect for Nature* that we can see each living organism as goal-oriented center of life, pursuing its own good in its own way. More precisely, he calls each living thing "a unified system of organized activity, the constant tendency which is to preserve its existence by protecting and promoting its well-being" (Taylor 1986, p. 45). He believes we should respect each organism's pursuit of its own good. In other words, for Taylor, all living things have equal moral status.

Although this position may seem to challenge our common moral intuitions, Taylor argues that it fits comfortably into what he calls a *biocentric outlook on nature*: the belief that humans along with other species should be seen as parts of the interdependent system of life on Earth. That is, humans are like other animals in that their survival depends on their relations to the physical environment and other living things. This outlook doesn't prove that humans and nonhuman beings are morally equal, but it does tend to undermine any argument for saying humans are inherently better or more valuable than other parts of the biotic community. Of course, humans are different in many ways from other living entities—but those entities are also very different from one another. Why should we regard humans as more important or morally worthy than the other members of Earth's community of life?

Of course, the belief that all living things have equal moral status leads to the difficult task of deciding when humans are justified in harming other lives, as we must do in order to live. Just breathing and walking around can inflict harm on the living organisms around us. If we embrace Taylor's view, we still need some way to decide what constitutes respecting an organism's life and what constitutes a good reason for harming it, and how to balance the interests of different organisms. And these considerations usually lead us back to qualities like sentience and rational capacity. Taylor himself, for example, acknowledges that because humans are capable of rational autonomy, they have a need for freedom and rights that animals and plants don't share. He also recognizes that we have greater duties to sentient animals than to plants.

Autonomy, sentience, and life aren't the only qualities that might be relevant to moral status, though. For example, the philosopher Clare Palmer points out that humans' duties to nonhuman animals may depend not only on the animal's capacities but on its relationship to humans. That is why we may feel greater obligation to our pets than we do to wild animals; pets depend on us in a way that wild animals do not. You may be able to come up with other qualities that seem important to thinking about our duties to nonhuman organisms. Perhaps we value their sociality, their capacity for joy, or their ability to create. How should we decide among all these possible reasons for recognizing duties to nonhumans?

The philosopher Mary Midgley suggests that we don't have to. She argues that there are actually many different bases for moral status. That is, some beings have moral status simply because they are alive; others have status because they are sentient, or have autonomy, or have a particular relationship to us. Midgley describes a complex moral world in which many different kinds of reasons can support our duties to different kinds of beings.

Midgley's theory is appealing because it describes pretty well how we actually conduct our ethical lives. That is, we often do give different reasons for respecting nonhuman lives. The chief objection to Midgley's approach is that she doesn't provide us with any way to decide how to weigh our duties toward all these beings against each other. For example, Midgley would tell Judy K. that she has a moral duty toward the deer because they are sentient and subjects of a life. But she also has a moral duty to her tenants; they are sentient and have rational autonomy, and she has a special relationship to them, since they depend on her to make good decisions about managing the property and keeping them safe. Similarly, Judy K. could also have duties to the trees and other plants in the forest, because they are alive, too. In short, Judy is surrounded by beings with moral status. How should she reconcile all these duties? If we had only one basis for moral status — sentience, for example — we could weigh duties according to how much pain or pleasure the being is capable of. Humans would get the most weight; deer would come in second, and the trees would probably not get any weight at all.

The single-standard approach is appealing because it promises to offer clear answers to hard questions like the one Judy faces. But it may achieve that clarity by ignoring morally relevant facts. Focusing only on sentience gives you an answer, but is it the right answer? Do we really believe that sentience is the only basis for moral status? Midgley's approach to moral status is more complicated, but maybe that's because the moral world we live in is complicated. Perhaps the value of Midgley's theory is not that it tells us the right answer but that it tells us why it's hard to find the right answer.

4.2.5 Pluralist-Expressivist Value Theory

Philosopher Elizabeth Anderson thinks all these approaches are too simplistic for the complex world we live in. She suggests that instead of just paying attention to how much we value things, we should think about *how* we value things. In her book *Value in Ethics and Economics*, she argues that different *modes of valuation*, like respect, awe, love, and admiration, are appropriate to different kinds of goods and beings. (This is why her theory is "pluralist": it recognizes many different modes of valuation.) While humans in general should be respected, for example, one's children should be loved. Similarly, some objects may be treated as commodities that can be owned, bought, and sold: Their proper mode of valuation is *use*, or subordinating the object to one's own ends. But the Grand Canyon should be treated differently, with admiration or awe. Your ethical duty is not simply to value these things enough but to value them in the right way.

How does this work? How do we know "the right way" to value something? Anderson insists that these different modes of valuation are not a matter of purely subjective, individual preference; they are a matter of social norms and practices. To begin, she claims that valuing something properly is not just a feeling; it involves conduct and expression. Therefore, an individual's ability to value something appropriately depends on the existence of shared norms, institutions, and practices that tell us how to act to express the proper attitude toward something. (This is why her theory is "expressivist": It considers valuing to be a matter of expression.)

For example, in American society, going to church and praying is a common and understandable way to value your deity appropriately, with reverence. Most people raised in the United States are familiar with this practice, even if they don't go to church themselves. This mode of valuation (reverence) is possible because we have the social practice that allows people to express reverence. Of course, many communities have different social practices that are used to express reverence: dancing, singing, bowing toward Mecca, fasting, etc. Because how one expresses reverence is a matter of social practice, it can vary from one community to another. Moreover, if you found yourself in a community that didn't have any set of practices around the value of spiritual reverence, it would be very difficult to express that value at all. Of course, you could still pray or bow toward Mecca, but it would be very difficult to explain to others what you were doing and why.

This is an important point: Anderson argues that being able to explain to others—and to yourself—what you're doing is what it means to value something rationally. In order to explain what you're doing, you have to make reference to some *social ideal*. In Anderson's words, "Ideals set the standards of conduct and emotion people expect themselves to satisfy with regard to other people, relationships, and things" (Anderson 1993, p. 6). We have been discussing reverence, which is one social ideal: a reverent person feels and appropriately expresses reverence for his or her deity. Similarly, a patriotic person expresses the social ideal of "patriotism" through certain conduct. In the United States, for example, one can express patriotism by saluting the flag and standing respectfully during the national anthem – or, for some athletes, by taking a knee during the national anthem. It is common for people to disagree about how best to express one's values, and what those values truly mean. But in order for people to have these arguments, they must share a basic understanding of the social ideal, its proper mode of valuation, and the social practices around it.

This theory is useful for making sense of many debates in environmental ethics. For example, Anderson suggests that the appropriate mode of valuation for most animals is consideration — that is, basic kindness. I would argue that many Americans believe we owe pets more than that, perhaps even family love. These modes of valuation (consideration and family love) express a social ideal we can call *animal friendship*. If we embrace that ideal, then we shouldn't treat animals as mere commodities, as though they exist just for our use and convenience. That is certainly not how we should treat our animal friends! Similarly, we shouldn't treat the Grand Canyon or an endangered wildflower as a mere commodity; the appropriate mode of valuation for such special parts of nature is awe, or reverence, or at least appreciation.

Anderson's theory gives us another way to understand some of the ethical arguments we discussed above. For example, we might say that Paul Taylor's book is offering "respect for nature" as a new social ideal. His book offers a philosophical argument but it also describes an ideal, a condition on Earth in which people are able to pursue their individual interests and cultural ways of life while at the same time allowing many biotic communities in a great variety of natural ecosystems to carry on their existence without interference. If we find that vision appealing, we should show "respect for nature" by conducting ourselves in the way he recommends, by refraining from harming wild plants and animals and limiting our impact on wild ecosystems.

Of course, we may worry that Anderson's theory says that how we ought to value animals (and other beings and goods) depends on social norms, so she sounds like a cultural relativist. And we know some difficulties with cultural relativism. We have already noted that members of the same culture may disagree about the social meaning of a good, and that can lead to conflict about what sort of behavior is rational. For example, there is much disagreement about the proper way to express consideration for animals: Some people believe that consideration for wild animals means they should never be subjected to human domination, so we shouldn't take care of them when they are injured. Others argue that taking care of an injured wild animal is a good expression of consideration. These groups disagree about what "animal friendship" really means and requires. And still others might reject the ideal of animal friendship altogether. Which of these positions is right, or at least well-justified?

Anderson explains that to justify an ideal like animal friendship, "one must be able to tell a story that makes sense of the ideal, that gives it some compelling point, that shows how the evaluative perspective it defines reveals defects, limitations, or insensitivities in the perspectives that reject these valuings" (Anderson 1993, p. 92). Therefore, in order to defend animal friendship, I would refer to the community's collection of stories about friendship between humans and animals, including wild animals. I would use these stories to demonstrate that this relationship offers emotional satisfactions unavailable to the non-animal lover. I might criticize those who don't share this ideal as cold, hard, and unhappy. I might offer empirical evidence for those interpretations. Note that under Anderson's approach, moral argument is not confined to debating whether animals can reason or suffer (although those points would probably come up). It also includes rich descriptions of what friendship with animals can be like. This method of moral reasoning looks very different from the method described in Sect. 2.3 as "the method of ethical inquiry."

Anderson's theory won't appeal to people who are looking for universal principles on which to rest environmental ethics (moral objectivists, for example). But recall the pragmatists' argument that the purpose of ethics is to help us navigate the moral world. Anderson's theory might help us understand and make effective ethical arguments in environmental matters, even if it doesn't rest on objective, universal principles.

Will this discussion help Judy K. navigate the ethical issues surrounding her deer management problem? At the very least, it suggests that the deer probably do have moral status based on the fact that they are sentient. If you agree with Tom Regan and Paul Taylor, you would recognize that they have a right to some degree of freedom to pursue their own lives. That doesn't mean that she has to let them destroy her garden and forest, though. These theories tell us that we should give the deer some consideration; but their interests must be balanced against the interests of other morally relevant beings. So we have only begun our analysis. Judy also has to consider her duty to the forest and the various species living there.

Is it Ethical to Eat Animals?

One of the most difficult questions in environmental ethics is whether respecting animals is consistent with eating them. On the face of it, ethical vegetarians have a strong case: Most people think killing a living being is harming it. Under any moral theory, if we are going to harm a being with moral status, we have to justify that harm. And our desire to eat meat doesn't seem like a very weighty justification. After all, most people in modern industrial societies don't need to eat meat; we have lots of other food choices.

One response to this argument is that killing an animal (humanely, without pain) isn't *really* harming it. Animals (under this view) don't fear death; they don't invest in the future. They will all die sometime, and dying in order to feed humans is no worse than dying of age or disease. Moreover, most of the animals we eat are domesticated varieties that wouldn't exist at all if we weren't raising them for food. Isn't having a short but happy life better than not existing at all?

Of course, this response assumes that our domestic animals do have a comfortable life. Most of them currently do not. For example, the vast majority of the meat consumed in the United States comes from animals raised in contained animal feeding operations (CAFOs). CAFOs have few defenders. But what about animals raised in more humane ways?

For example, consider the owner of a bison ranch in South Dakota. His bison roam the native prairie much as their ancestors did. When he receives an order for bison meat, he carefully selects and kills one of the herd. This culling helps him maintain a healthy herd that won't overgraze the prairie. The prairie also benefits as an ecosystem from the bisons' grazing. Is it unethical to eat this rancher's bison burgers?

It's a tricky question. The bison will of course die one way or another, and their deaths from age or disease might be more painful than the death the rancher inflicts with his rifle. We may also worry that without the market for bison burgers, the whole species of bison may disappear. Finally, there's the fact that producing crops also kills millions of animals, either directly, in the course of plowing and harvesting, or indirectly, through its impact on habitat. Traditional moral philosophy does not offer neat solutions to this ethical dilemma.

Elizabeth Anderson's value theory offers a different way to approach the question. Under Anderson's theory, the relevant question is, "What does it mean to raise animals for food?" That is, is this practice consistent with our ideals of animal friendship? Some would argue that to kill an animal is to betray your friendship with it. But hunters and farmers would argue that their practices express good, meaningful ideals. Hunters argue that they show respect for the prey animals by following the code of the hunt. Farmers argue that they show consideration to their animals by respecting the "domesticated animal contract": the implicit agreement to give their animals a good life and a humane death. Under Anderson's view, ethical argument on this question should not focus simply on whether death is a harm; it should explore the meaning and value of these ways of living with and from animals.

4.2.6 Species

We may be persuaded that we have duties to individual organisms, or at least individual animals. But do we have duties towards species as a whole? This question is more difficult to answer. True, the United States and many other countries have enacted laws to protect endangered species, even at the expense of economic development. A lot of people accept the principle that we have a duty to preserve species. But a duty to whom – to other humans, or to species themselves? Are species the kind of thing we can have duties toward? That is, do species have moral status?

It helps to understand what we mean by *species*. "Species" is a term most commonly used by biologists to categorize organisms. For biologists, a species is basically a group of organisms that are genetically similar. How similar? As with any classification system, it depends on your purpose. For example, people who study bacteria often say that the concept of species isn't useful to them at all, because genes move so rapidly among bacteria. But ecologists do find it useful to classify organisms by species. Genetic similarity is often the result of adaptation to a common ecosystem, and that's what ecologists are interested in. But similarity might also be the result of a breeding program. Lab mice, for example, may be genetically identical. Does that make them a distinct species? If so, are they as worthy of protection as endangered eagles or polar bears?

Some biologists define species as a community of organisms that can interbreed and produce viable offspring. This is a good working definition, although it doesn't work for the huge number of species that reproduce asexually. And even among organisms that reproduce sexually, there is a lot of gray area; populations of organisms are often in the process of speciation. At what point are they different enough to be designated a separate species?

Ultimately, *how* we define "species" depends on *why* we are defining "species" . That means we have to decide why species have moral value before we can decide what counts (ethically) as a species. This is an important point that will keep coming up: Environmental ethics necessarily borrows a lot of concepts and terms from the natural sciences, like "species," "ecosystem," and "biodiversity." It's tempting to think that these terms, being scientific, have clear, objective, value-neutral, uncontroversial meanings. But they don't. Scientists themselves debate their meaning and even their usefulness. How we define these terms usually ends up resting on a value judgment: We define "species," for example, in a way that reflects something the researcher or society values about the natural world. It's important when borrowing a concept from another discipline to become familiar with the debates and controversies over its meaning, and to dig down beneath the façade of objectivity to the value judgments that these concepts reflect.

That point brings us to the next problem. We have discussed several bases for recognizing the moral status of animals. Individual animals may have desires, beliefs, and goals; they may experience pain and pleasure. We can have duties towards them because they have *interests*. A species, in contrast, has no desires, beliefs, or goals. A species is just a category, like "people with red hair." While we certainly have duties to individuals with red hair, it's hard to see how we could have a duty to the *category* of people with red hair. Similarly, while we may have duties to individual members of a species, it's hard to explain why we have a duty to preserve the species as a group. At any rate, the explanation will have to be different from the explanation for our duty to individual animals.

One possibility is that the duty to preserve species is simply an extension of our duties to individual animals. Some animals are highly social and can only thrive when they live in community with their conspecifics. Horses and primates are examples. If we want horses and primates to live good lives, we have to ensure that they have the right sort of company, which means we should ensure the continuation of others of their kind. That reasoning, however, doesn't explain why we seek to preserve species that aren't social in this way, or why we want to preserve plants (which presumably don't care about their conspecifics).

A more common argument is that we preserve species not for the sake of the species themselves but for our sakes. That is, species have no intrinsic value, but they have instrumental value to humans. It's good to live in a world with lots of different species. (Or maybe it's just good to live in a world with lots of naturally-occurring, non-engineered species?) We may worry that the loss of some plant species might deprive us of a cure for a deadly disease, or that a species might play a critical role in an important ecosystem. This is a good argument for preserving some species, but not all existing species offer this sort of practical value to humans. True, the loss of any species deprives us of the opportunity to study it and learn more about the processes of evolution and adaptation. But we can't be sure that the species we lost won't be replaced by another, equally interesting or valuable species. So why do we feel obligated to preserve the *existing set* of species?

Lilly-Marlene Russow has made the interesting argument that the duty to preserve species is ultimately grounded on a special kind of instrumental value: aesthetic value. This argument requires a little digression, because academic philosophers usually treat aesthetics as an entirely different subject than ethics. Traditionally, aesthetics concerns how we make judgments about beauty and taste, while ethics concerns how we make judgments about right and wrong. But if we think of ethics as concerning how to live a good, choiceworthy life, then aesthetics are certainly an important part of ethics.

We'll discuss aesthetics in more depth in Sect. 7.5. For now, the important point is that aesthetic qualities are qualities of an object or experience that contribute to (or detract from) one's appreciation of it. They are the qualities that make something beautiful, mundane, boring, disgusting, delightful, intriguing, etc. These include sensory qualities: how it looks, sounds, smells, feels. But they can also include less tangible qualities, like complexity, rarity, and unexpectedness. Aesthetic value might also depend on the things you know about the object, such as its history and meaning.

Russow's argument is that we want to preserve eagles, polar bears, primates, and even modest little plants like the endangered dwarf trout-lily because members of those species have aesthetic properties we value. They may be beautiful, majestic, cute, complex, elegantly simple, fierce, endearing, unexpected, amusing; there are quite a variety of aesthetic values that we find in the nonhuman world. This reasoning explains why we value rare species more than common ones: Rarity is an aesthetic value in itself, and rare species are in danger of becoming extinct, which would mean the loss of all that species' aesthetic value. This theory also explains why we mourn the loss of a species even if another one develops to fill its place. The other species may also have aesthetic value, but it will not exhibit the *same* aesthetic properties. In the same way, we would mourn the loss of a painting by Picasso even if it were replaced by a Renoir. The Picasso has distinctive qualities that we don't want to lose.

Of course, that analogy is a little misleading. The reality is that we don't think the species we are losing will be replaced any time soon. Many ecologists argue that human modification of ecosystems in the name of economic development has created a mass extinction event: the relatively sudden and dramatic decline in species diversity all over the world. So we aren't just ending up with a different collection of species, as though we traded our Picassos for some Renoirs. We're ending up with fewer species altogether, as though the entire world art collection were decimated. This loss of value is staggering, and efforts to preserve species are surely an appropriate response to this loss.

But here's the complication: If we want to preserve species for their instrumental value — because they make human life better — then we have to weigh their loss against the instrumental value of all things we *gain* through economic development: electricity, health care, adequate nutrition, etc. These things also make human life better. Perhaps some trade-off between species richness and other goods is acceptable.

Still, it seems wrong to say that more health care can always compensate for the loss of polar bears and bald eagles. As Russow's theory makes clear, these goods have different *kinds* of value, and surely we want a world that is rich in many different kinds of goods and experiences. Elizabeth Anderson might say that *species richness* (or biodiversity, a related concept) is an ideal, just as animal friendship is the ideal underlying our animal welfare ethic. Using Anderson's term, the proper mode of valuation for species is *appreciation*: appreciation for their beauty, complexity, uniqueness, quirkiness, and other aesthetic properties. Our endangered species policies are justified to the extent they properly express that value. We appreciate species by trying the preserve them, even in the face of pressures toward economic development.

4.2.7 Ecosystems

So far we've considered Judy K.'s duty to individual deer and to the species that occur on her land. Her ethical problem grows much more complicated, however, when we consider her duty to the forest. Is a forest the sort of thing one can have a duty toward? If not, what exactly is her duty with respect to the forest ecosystem?

An ecosystem, like a species, is a term invented by biologists to categorize the world (and we know that we have to be careful when we borrow scientific terms!) Ecosystems, like species, lack beliefs, desires, and interests. In fact, the definition of "ecosystem" is even looser than the definition of "species." An ecosystem is, as the term suggests, a *system*. More specifically, ecologists usually define "ecosystem" as a network of functional interactions among organisms and between organisms and their environment. But one can define ecosystems at different geographic scales. That is, one might be interested in the whole forest ecosystem, which could cover several hundred acres, or one might be interested in a small pond in the forest, covering only half an acre. Or one might be interested in the area of transition between forest and grassland, which could constitute an ecosystem in itself. And to make things even more confusing, ecosystems (like species) are always in the process of changing. All of this makes it difficult to decide what counts as *harming* an ecosystem. If you fill in the pond in order to help the forest thrive, have you destroyed an ecosystem or helped it? What if the pond was in the process of turning into a meadow; haven't you simply helped along this natural transition? (This point can be reframed as a problem of temporal scale: Are we interested in the ecological processes that we can see over the course of days or months, or the ones that unfold over many years or decades?)

As with species, in order to determine what counts as an ecosystem for ethical purposes, we have to decide what it is about ecosystems that we value. Here, again, the instrumental value of ecosystems seems most important. We value ecosystems in large part because these systems, or communities, are necessary to the individual organisms within them. That is, our duties to ecosystems may simply be an extension of our duty to preserve and attend to the welfare of individual organisms. We want to preserve the forest because we value the plants and animals that make up the forest ecosystem.

But this duty to organisms doesn't seem to capture everything we value about ecosystems. Just as we appreciate species richness (a wide variety of species), we also value ecosystem richness. Ecosystems like organisms have aesthetic properties: beauty, complexity, grandeur, simplicity, etc. We may want to preserve a variety of different kinds of ecosystems for the same reasons that we want to preserve a variety of species. Ecosystems, like species, should be appreciated and cherished for themselves, not just for the sake of the organisms that depend on them.

But preserving ecosystems poses a challenge: Many ecosystems require active management by humans. One difficulty we encounter here goes back to geographic scale: The scale of the ecosystem might not match the scale of the moral community or political organization we've created to manage it. A desert, for example, might be bigger than any single political jurisdiction, and some ecosystems (like the Arctic or the world's oceans) might not be under any particular government. But even if we could fix that problem, we have a deeper problem. Managing ecosystems might require us to kill individual animals. Judy K., for example, may need to kill some deer to preserve the forest. Is this justified? How do we weigh our duty to attend to the welfare of individual organisms against our duty to preserve an ecosystem? And what if preserving an ecosystem requires us to eradicate a species? How do we balance those values?

This is a topic of considerable debate in environmental ethics. Some environmental ethicists argue that we should always value the system over its individual parts; they contend that ecosystems are always more valuable than individual animals. Environmental philosophers call this a *holistic* ethic: The whole is more important than the parts that make it up. We can contrast holistic ethics with *reductionist* ethics, which focus on the value of the parts that make up the system. We considered above the reductionist argument that the system is valuable in large part *because* the individual organisms in the system depend on it. That is, the individual organisms have value in themselves, not just because they are parts of the system.

If you embrace a holistic ethic and value the ecosystem over its individual parts, you might conclude that some of the deer should be removed to protect the forest. But even if you reject this holistic view, you might still conclude that we must get rid of some of the deer to make sure that the rest of the deer thrive. That is, we could justify hunting deer in utilitarian terms: The harm to a few deer is outweighed by the benefit to all the other organisms in the forest. That's a compelling argument, but we must ask ourselves why we don't pursue this logic when it comes to humans. That is, we don't simply kill off humans in order to preserve the ecosystem they (and other organisms) depend on. Is this because humans are more valuable than deer? We discussed above some reasons for thinking they are: unlike most other animals, they enjoy rational autonomy. Perhaps that value outweighs all the value of other organisms that we sacrifice for the sake of humans. Or is it better to say, with Elizabeth Anderson, that we should value humans in *a different way* than we value deer? Perhaps hunting deer is consistent with our ideals of animal friendship and consideration. Hunting humans, in contrast, is treating them as though they are deer, which is a moral mistake. Other social ideals apply to our relationships with humans.

4.3 The Deer Problem Revisited

This discussion suggests that there may be no simple, standard answer to Judy's deer dilemma. Making decisions about environmental management often requires us to balance duties to animals, species, and ecosystems. We want to show consideration to animals and attend to their welfare; but we also want to preserve a reasonable degree of species and ecosystem diversity, both for the sake of the organisms themselves and for the sake of living in a world filled with a rich variety of beings and places. A good decision will attend to all these values, balancing them (as always) against the value of other things we need to live good, choiceworthy lives. For Judy K., that might mean instituting a limited hunt to keep the deer population under control, in order to protect the forest and her garden. She would have to consider the most humane way to kill the deer, as well as the potential danger of the hunt to humans. The best solution might be to allow only the most skilled bow-hunters to participate in the hunt; bow-hunting poses less threat to humans and a good bow-hunter can kill a deer quickly and cleanly. Moreover, killing just a few deer each year might be enough to encourage them to forage more widely, which would be better for the deer as well as the forest. But other situations may call for a different calculus: If the deer were particularly rare, or the forest ecosystem more delicate, or the garden of great historical importance, Judy might reasonably adopt a different solution.

The deer problem may seem pretty complicated. But we haven't yet considered everyone who has an interest in Judy's decision. The next chapter expands the temporal scale of ethical decision making, exploring how we should take into consideration the interests of future generations.

Further Reading

Anderson, E.: Value in Ethics and Economics. Harvard University Press, Cambridge (1993)
Frey, R.G.: Rights, interests, desires and beliefs. Am. Philos. Q. **16**(3), 233–239 (July 1979)
Katz, E.: Is there a place for animals in the moral consideration of nature? In: Light, A., Rolston III, H. (eds.) Environmental Ethics: An Anthology, pp. 85–94. Blackwell Pub, Madden, MA (2003)
Midgley, M.: Animals and Why They Matter. University of Georgia Press, Athens (1984)
Palmer, C.: Animal Ethics in Context. Columbia University Press, New York (2010)
Regan, T.: The Case for Animal Rights. University of California Press, Berkeley (1983)
Rolston III, H.: Duties to ecosystems. In: Callicott, B. (ed.) A Companion to the Sand County Almanac, pp. 246–274. University of Wisconsin Press, Madison (1987)
Russow, L.-M.: Why do species matter? Environ. Ethics. **3**(2), 101–112 (1981)
Singer, P.: Animal Liberation. Avon Books, New York (1977)
Sober, E.: Philosophical problems for environmentalism. In: Norton, B. (ed.) The Preservation of Species. Princeton University Press, Princeton (1986)
Taylor, P.: Respect for Nature. Princeton University Press, Princeton (1986)
Walker, A.: In Search of Our Mothers' Gardens. Harcourt Brace Jovanovich, New York (1983)

Chapter 5
Do We Have Duties to Future Generations?

Contents

5.1 The Future of Spring Lakes

Decisions about land management often have effects very far into the future. Land that is mismanaged might take decades to recover. If a rare species goes extinct, future generations will never experience it. Therefore, the policy makers and other decision makers who deal with environmental management often use a *planning horizon* of several decades or even a couple of centuries (a time scale that is much bigger than most ecological research, by the way — which means there's often a mismatch between the temporal scales of ecological science and land use planning!)

Judy K.'s planning horizon extends for several decades; she thinks about her children and grandchildren when she makes decisions about Spring Lakes. That's natural. Her children will likely inherit the land, so they have an interest in how it's managed. She assumes that her grandchildren and their children will also be interested in keeping Spring Lakes in good ecological health. In other words, she acts as though she has a duty to future humans who will enjoy her land. Like many land managers, she is thinking several generations ahead.

Should she? The notion that we have duties to future generations is widespread; indeed, most people take it for granted. But some philosophers have pointed out conceptual difficulties with this idea. Their arguments can help us better understand exactly what sort of duties to the future we might have and how to balance them against our duties to the current generation. In other words, these philosophers are thinking about the challenge of changing the traditional temporal scale of ethics to fit the long-term planning horizon that sustainable environmental management requires.

K. K. Smith, *Exploring Environmental Ethics*,
AESS Interdisciplinary Environmental Studies and Sciences Series,
https://doi.org/10.1007/978-3-319-77395-7_5

5.2 Future Generations

In his book *Reasons and Persons*, the philosopher Derek Parfit explains one problem with the notion of duties to future generations: Any group of humans in the future will owe their existence to us. Whatever we, the current generation, decides on any given environmental issue (or any other policy question) will affect the identity of the next generation. Even the smallest decision, like whether to take a vacation in California or Florida, can affect when a person is conceived and therefore the genetic endowment of that person and the context into which she or he enters the world. In short, *our* decisions bring this particular set of humans into existence. So how can they complain of the state of the world we leave for them? If we had made different decisions, they wouldn't exist — some other group of humans would have been born. (This is often called the *Nonidentity Problem*, because the persons we think we're harming with a bad decision are not the same as those who would benefit from a good decision.)

For example, what if Judy K. allows her son to hunt deer for a couple of weeks this year, in order to control the deer population? It's possible that if her son hadn't come to visit for those 2 weeks, he and his wife might have conceived another child during that time. That child will now never come into existence. If Judy instead let the deer ravage the forest, the child would exist but would have a somewhat less healthy forest to inherit. Should the child complain? Isn't it better to exist even in an ecologically diminished world than not to exist at all?

This logic may seem like quibbling. Even Parfit concludes that his logic goes against widely shared moral intuitions and is not a good guide to public policy. He suggests that we shouldn't worry too much about the Nonidentity Problem and simply try to bring the best possible future world into existence.

But that approach runs up against another conceptual problem. What sort of world will future generations want? To be sure, future generations will have some basic needs common to all humans: breathable air, clean water, a functioning agricultural system. Some philosophers argue that we can go a bit further than that. Richard Vernon, for example, has argued that we can think of future generations as being similar to citizens of other countries: They may not be members of our political community, but they are vulnerable to our actions. Following the reasoning offered by Robert Goodin and Andrew Dobson (discussed in Sect. 3.7), this condition of vulnerability means that we have a duty as good citizens to consider their interests in making decisions about public policy. Vernon concludes that we can at least try to pass on to future generations the conditions that make it possible for them to enjoy just political relations with one another. That might require more than a healthy environment; perhaps we should also make sure they have institutions, knowledge, and traditions that support justice.

But can we make any further assumptions about what these future generations will value? Judy assumes her grandchildren and great grandchildren will value a hilly, forested bit of mid-Michigan. But by the time they're born, social values may have changed. Maybe they would be happier if great grandmother Judy had sold the land to a developer and invested in tech stocks. Once we start making guesses about what people will need 50 or 100 years in the future, the uncertainties multiply. So it may be unwise to give these *far future duties* much weight — especially against the more pressing, obvious, and undeniable duties we have to the current generation.

5.3 Future Generations and Public Policy

The uncertainty of the future has been particularly troublesome for environmental policy. Consider the case of climate change: Because greenhouse gases remain in the atmosphere for a very long time, any effective policy to combat climate change has to have a very long planning horizon. The decisions we make today will affect generations that exist 100 or 200 years from now. Some climate policy experts argue that because of these uncertainties, we shouldn't put much weight on our duties to those future generations; that is, we should focus on growing our wealth and take care of existing humans. That logic may get some implicit support from Parfit's Nonidentity Problem: After all, even if we give future generations a world with an unstable climate, surely that's better than not existing at all, isn't it?

The British economist Nicholas Stern has criticized that position. He argues that we have roughly the same duties to all humans, existing and future. Justice is treating equals equally, and we have no good reason to treat future generations differently than currently existing ones. In his view, giving less weight to the interests of future humans is simply a *presentist bias*; it's like favoring your friends and neighbors over people who live farther away. (It might be ethical for private citizens to favor their friends, but policy makers are supposed to be more impartial.) Stern also thinks that we can estimate the likely effects of climate change with enough confidence to know that a cooler world will be substantially better for the future than a warmer one. Stern's analysis concludes that we should be willing to sacrifice about 1% of the world's gross domestic product each year, in order to keep global mean temperatures from increasing by more than 5 degrees Celsius. (The world GDP was about $50 trillion when he published this argument, so he was suggesting that we should spend $0.5 trillion annually on carbon emission reduction.)[1] Of course, Stern understands that this is a lot of money to spend on people who won't exist for another hundred years. Therefore, he also argues that the money we spend on developing cleaner, renewable, and more efficient technologies will have additional benefits for human welfare in the near future. Therefore, even if we're wrong about the harms created by climate change, we won't have reason to regret having chosen this greener development path.

Stern and Parfit both recognize that regardless of the conceptual problems, we're unlikely to convince people like Judy K. that she should stop worrying about future generations. The idea that we're bequeathing a good world to the future helps to satisfy our need for a meaningful life; this is one way to make our lives matter.

[1] Nicholas Stern (2010).

On the other hand, giving the interests of all future generations the same weight as those of the current generation probably isn't sensible. We don't know very much about the far future, but we do know a great deal about the people suffering right now. The challenge is to find a path forward that satisfies our duties to the present and leaves open plenty of options for future generations. The following sections explore that challenge in the context of two important issues for global environmental governance: setting the discount rate for climate policy and addressing overpopulation.

5.3.1 Ethics and Discounting in Climate Policy

Economists often have to consider how to allocate wealth between the present and the future generations. This topic comes up most often in economics in the discussion of discounting and the social discount rate. Examining this topic is a good way to bring two important disciplines — economics and ethics — into conversation with each other. This topic is an excellent example of why interdisciplinary training is valuable for people concerned about the environment.

When economists do cost-benefit analysis (weighing the costs of an action against its benefits), it makes sense to discount future benefits: While it might seem that $10,000 in 10 years from now is worth the same as $10,000 right now, it isn't. If you had $10,000 and invested it wisely, you would end up with more than $10,000 in 10 years. (The exact amount would depend on what sort of returns you could get.) In other words, $10,000 in 10 years is worth less than $10,000 right now. To capture that reality, economists usually discount future benefits, using an estimate of interest rates (a general measure of how much invested wealth can be expected to grow each year) to decide how much that *discount rate* should be.

Governments do the same thing. When deciding whether to spend money on a public project, they consider what else they could do with the money. One important question they must consider is what opportunities they are foregoing in order to fund this project. They factor those *opportunity costs* into their calculations, and they won't go forward with the project unless its benefits outweigh these costs. In other words, they discount the benefits by some factor related to what sort of return they could get on the money if it were invested somewhere else. This is called the *social discount rate*. Usually the social discount rate in developed countries is between 3% and 7%.

Climate policy experts have debated what discount rate to use when calculating the costs and benefits of combating climate change. While most people use a standard social discount rate of around 3%, Nicholas Stern's argument, discussed above, is based on the claim that it should be lower.

First, he points out that this is not a typical cost-benefit analysis. In most cost-benefit analyses, we can make assumptions about the world's general path of development; we're just choosing among different options within that path. But with climate change, we're choosing *different* paths of development. In fact, we are choosing between faster or slower world economic growth in the near future: either strong regulations on greenhouse gas emissions and slower economic growth, or lax greenhouse gas emissions and faster economic growth. That means we are choosing whether a higher or lower interest rate will be available for investors. So we can't simply assume a standard interest rate as the basis for our discounting. As Stern puts it, the interest rate is *endogenous* (meaning that it's an outcome of the model, not an independent input). An economist might put it this way: This is not an appropriate situation for *marginal analysis* because we are not making decisions "at the margin" but deciding between different economic systems altogether.

In this situation, we are clearly making trade-offs between generations: Should we accept a slower rate of economic growth for the next several decades in order to protect future generations from the losses caused by an unstable climate? Stern considers two reasons to favor the present over future generations:

First, one could engage in *pure time discounting* — that is, favoring the present just because they exist. Stern doesn't think there's much reason to do this. He does consider the argument that we are justified in favoring people close to us — our families and neighbors — over strangers. As we discussed in Sect. 4.2.4, some philosophers argue that having a special relationship to someone can be a basis for moral status. If you think that's true, that might be a reason to favor the present generation over future strangers. But Stern points out that such favoritism isn't appropriate for policy makers. As a mother you can favor your own child; as a government official, you have to treat friends and strangers the same. (This point is an example of the principle we discussed in Sect. 3.8, that one's moral obligations may be shaped by one's role.) Nevertheless, Stern does accept one reason for giving the present generation slightly more weight than future generations: There's a small risk future generations won't materialize. The world might be hit by a meteor that would destroy all human life, for example. That is a reason to use a very small (0.01%) pure time discount rate.

Second, Stern considers the argument that people in the future will be wealthier than we are, so they will value each additional unit of consumption less. That is, we normally think that $1000 is worth more to a poor person than to a rich person, because the rich person already has most of his needs met. So if we expect future generations to be wealthier than we are, we should favor the current generation. Stern agrees with this logic, up to a point. But he points out that people have different intuitions about how much we should favor the poor over the rich. Therefore, we ordinarily keep this discount factor fairly small so as not to disfavor the rich (in this case, the future) too much. He's also not sure future generations will be all that rich, at least in terms of environmental goods. They may be a lot poorer. In the end, he's willing to use a slightly higher discount rate to take into account the possibility that future generations will be richer than we are.

Ultimately, though, Stern warns us against focusing too much on the technicalities of discounting when deciding what to do about climate change. The judgment we're facing is quite complex: Choosing a greener path of development will have costs, but it could have a lot of unexpected benefits. And this choice also involves questions about risk, which are also value judgments: How much risk of economic recession are we willing to tolerate? How much would we pay to reduce risks to future generations? How should we handle the small possibility of a catastrophic climate change? There is very little social consensus on these questions, because people have different tolerances for risk. He therefore suggests a holistic decision making approach. We shouldn't just weigh the costs of action against the costs of inaction. We should also sketch out several plausible paths to the future and consider how much each path reduces the risks of catastrophe, and whether it increases future options for economic development. In addition, we should consider whether our estimates of costs and benefits are likely to be more reliable under some paths than they are under others (that is, how much uncertainty is inherent in each path). Not surprisingly, Stern concludes that transitioning away from fossil fuels and toward technologies that reduce greenhouse gas emissions is a better choice for the present and the future, all things considered.

Economic Value

Economists like ethicists are interested in what people value and how much they value it. But economists focus on only one dimension or kind of value: economic value. Goods and services have this kind of value if humans actually *do* value it, and express that value by paying (or foregoing) money for it. Economists therefore approach questions of value empirically (like anthropologists) rather than asking what we *should* value. The discipline of economics has developed very sophisticated measures of economic value, along with theories that explain how to maximize that kind of value.

Economists are aware that many things we value are not bought and sold in a market. Money can't buy you love — or a good character, a happy family, or a meaningful life. But a surprising number of things can be turned into market goods. A pollution permitting system, for example, can give you the right to put pollution into the environment. Allowing people to buy and sell permits creates a market in the right to pollute.

But should we create such markets? Economists can tell you how a market will affect economic value; for example, their models might show us that we can achieve more pollution control at a lower price by creating a market for pollution permits. But they can't address whether we *should* try to increase wealth in this way. We might worry that this sort of system could have a bad influence on our character, by teaching us to see pollution as a reasonable cost of doing business instead of an assault on the biotic community. Economic analysis also typically ignores the value of goods to nonhumans. Because animals don't buy and sell things, economists have no way to take into consideration how much they might value clean water or clean air. It's important to understand these limitations when relying on economic analysis.

5.3.2 Population Control

Overpopulation is also a special ethical issue related to the question of duties to future generations. Currently there are about 7 billion people on the planet. Demographers predict that if current trends continue, the human population will stabilize at around 10 or 11 billion by the twenty-second century. Many ecologists believe that 10 billion people would exceed the planet's ecological limits; in fact, at 7 billion, we may already be exceeding our limits. They worry that unless we halt and reverse population growth, future generations will inherit a world that can't support them all. This worry is not shared by everyone, of course; there is considerable uncertainty about how many humans the planet can adequately support. Perhaps technological advances and smarter land-use planning could lighten the burden of the human population. But if there is some upper limit to the planet's capacity, then we have to grapple with the question of whether there is an ethical approach to limiting the human population.

The ethics of population control pose some interesting puzzles. One question, of course, is discussed above: since future generations wouldn't exist at all if not for our decisions, do they have a right to complain about the conditions of their existence? How good a world do we owe them? A second, related question was explained by Derek Parfit, in the same book that introduced the Nonidentity Problem. Parfit is a utilitarian, so he wants to increase human happiness or welfare. But, he asks, should we aim to increase total happiness, or just average happiness? If we aim to increase total happiness, we should keep growing the population, adding more and more people, even though they will be less and less happy. The sum total of their happiness will still increase, even though each of them might be just slightly better than miserable. That doesn't seem like a good goal (Parfit calls it "the repugnant conclusion"). On the other hand, we could try to increase average happiness by reducing the total number of people but making sure they're all really, really happy. That might sound like a good idea, but can we justify telling a really happy couple that they shouldn't have a third child just because that child will be slightly less happy (but still have a very good life)?

There doesn't seem to be a good way to resolve this puzzle using utilitarian reasoning. Indeed, it had become common in discussions of population policy to approach the question from a different angle — from a rights-based framework instead of a utilitarian framework. Considerable empirical research now supports the position that when women have access to education, economic opportunity, and birth control, birth rates decline. Therefore, instead of focusing on how big the global population should be allowed to grow, many environmental policy experts argue that we should focus on protecting the rights of women. There's good reason to believe that ensuring women have the same level of autonomy as men will lead to a smaller and more stable population. In fact, it should also help to end hunger and malnutrition, since similar research shows that when women control household resources, food is distributed more fairly and children are less likely to go hungry. Indeed, it may be that solving many global environmental problems requires improving the status of women (which supports the ecofeminist position that the oppression of women and the overexploitation of nature are interdependent.) This focus on women's status has the virtue of supporting policies that we know how to implement (like making education and birth control available) and that show respect for human rights. In contrast, policies that aim to limit family size directly (such as China's attempts to lower the birth rate) have been criticized as intruding too much into intimate family decisions and having other undesirable social impacts.

The link between women's status and population suggests that as a citizen, the best way to address the overpopulation problem may be to support policies aimed at giving women equal educational and economic opportunity and full control over their reproductive decisions. The harder ethical question is how the risk of overpopulation should affect *your* individual decision whether to have a child. This is a matter of private ethics, not public policy. That is, one might believe that having a child is unethical (given the population burden on the planet) but also that the government should not interfere with a person's decision to have a child (because a liberal government has to respect a person's autonomy, her freedom to make an unethical choice).

If one believes that we are approaching the ecological limits of the planet, then it seems hard to justify bringing another person into the world, particularly when adoption is a viable option. Adoption allows one to achieve the goods of parenthood without adding to the ecological costs. True, some people argue that adoptive parenthood is inferior to "true" parenthood, but that position is hard to support philosophically. The stronger objection is that the legal system might make adoption very difficult for some people, but that is a problem that could be solved with better laws. If adoption is relatively easy, why would one choose to bring another child onto a crowded planet?

On the other hand, the notion that having a child is unethical is very hard for most people to accept. Most religious and cultural traditions teach that having a child is a blessing and perhaps even a duty. Indeed, for most of human history, societies have benefitted from population growth. Even now, some political leaders worry that population decline in their own countries will impose social and economic costs. (However, others reject this "pro-natalist" position on the grounds that those costs can be avoided by allowing more immigration.) Surely we want *some* people to reproduce so that we can maintain a viable human civilization, and surely no one wants children to be born into a society that views them as a burden and a curse. But it seems wrong to view having a child as a special privilege to be confined to people who have some special sort of merit or resources.

Like most questions in environmental ethics, there is no universally accepted answer to this puzzle. But I think we can draw a few conclusions from this discussion. At the very least, we may conclude that choosing to limit one's family size or not to have children at all can be justified. There is at least some reason to believe that there are ecological limits to human population growth. Unless we achieve greater certainty that the planet can support the current population without degrading our ecological support systems, we shouldn't pressure people to have children. We may also conclude that one should at least make sure one's own children have a good life, and aim to have no more children than one can support. These commitments, along with support for policies that protect reproductive freedom, should help to create a world in which we are not faced with the tragic choice of either protecting the planet or having children. Whether or not we can say that this is a *duty* we owe to future generations, we can at least say that such a world would be good to live in, and it would be good for us to bring it into being.

5.4 The Future of Spring Lakes Revisited

What does this discussion of future generations mean for an ordinary decision maker, like Judy K.? One important lesson she might take is that she shouldn't be too confident that she knows what her own descendants will want or need. In fact, it may be that her grandchildren will decide not have children themselves (a choice that she should probably support), or will choose to sell Spring Lakes to others. These potential future owners might have very different values and interests than Judy has. It is a disconcerting thought: What sort of obligations should she feel to these strangers, people she has never met or imagined? Isn't it foolish to think that she can manage the land in their interests, when she has no idea what those interests will be?

But we are not completely without guidance in planning for the far future. At the very least, Judy could take Nicholas Stern's advice and attempt to manage Spring Lakes by mapping out several different paths towards the future. For example, she might consider a plan that focuses on preserving or restoring native habitat and cultivating biodiversity versus a plan that would add more housing and make the place more profitable. Which plan would leave more options open to future owners of the land? Perhaps she would conclude that ecological preservation and restoration would be appreciated by herself and her children, and wouldn't prevent future generations from using the land in a different way if they needed to. Or she may discover some opportunities to add housing in a way that wouldn't diminish biodiversity and would strengthen the community, giving her and her family more resources to take better care of both the land. She can at least attempt to choose a path that future owners would be unlikely to regret.

Another way to think about this question of duties to future generations is to say that Judy K. is trying to endow her land with a *good history*. That is, her aim is to give future generations a good story about what she did and why she did it. There's no guarantee that future generations will share all of her values. But acting thoughtfully and giving careful attention to the variety of beings and interests that make up the current biotic community should go a long way toward ensuring that the future Spring Lakes has a history of being loved and cared for.

Further Reading

Golding, M.: Obligations to future generations. Monist. **56**(1), 88–95 (1972)
Stern, N.: The economics of climate change. In: Climate Ethics: Essential Readings. Oxford University Press, Oxford (2010)
Parfit, D.: Reasons and Persons. Oxford University Press, Oxford (1984)
Vernon, R.: Justice Back and Forth. University of Toronto Press, Toronto (2016)

Chapter 6
Property and Stewardship

Contents

6.1 The Problem of Sand Mining

Judy's neighbor Maria called her the other day to tell her about a new threat to the little community at Spring Lakes: sand mining. The growth of hydraulic fracturing (fracking) has created a demand for the kind of sand used in the process. A local company has requested a permit from the county to mine sand from a lot just a few hundred yards from Judy's property line. Maria is outraged and wants to organize opposition to the permit. Judy, as usual, isn't sure she agrees with Maria. On one hand, the mining will create some noise and truck traffic on the road. But she's used to that; there used to be a gravel mining operation right down the road from Spring Lakes, and that wasn't too disruptive. Like Maria, she worries about whether the mining would have negative ecological impacts. On the other hand, she understands why the property owner wants to make some money off of his land. Judy also is thinking about letting a timber company to remove some aging oak trees from Spring Lakes. The money from this operation would help her maintain the land in better condition. She certainly doesn't want her neighbors interfering with that plan. So does she have any right to interfere with her neighbor's sand mining operation? What is her duty here?

K. K. Smith, *Exploring Environmental Ethics*,
AESS Interdisciplinary Environmental Studies and Sciences Series,
https://doi.org/10.1007/978-3-319-77395-7_6

6.2 Property Rights

Judy's dilemma arises out of the potential conflict between private property rights and the community's interest in environmental goods. This is a very familiar problem in environmental policy, since most environmental policies involve putting restrictions on what private citizens can do with their property. The United States has a tradition of giving strong protections to private property rights, so any new restrictions on property rights are generally met with resistance. Justifying environmental protections requires us to understand the ethical reasons for protecting private property rights as well as the ethical reasons for restricting them.

6.2.1 What Is a Property Right?

Let's begin with the basics: What exactly is a property right? Again we are using a term borrowed from another discipline: law. So we should begin by considering how legal scholars define the term. Legal scholars teach us that a property right is really a cluster of rights: When you acquire a right to a piece of property, you may acquire the right to enter the property, use it, change it, sell it, rent it to others, lend it, give it away, or even destroy it. But the law can limit those rights in various ways, giving rise to different property systems. For example, in one country, owning a piece of land might mean that you can sell it or bequeath it to anyone you want; in another country, those freedoms could be severely restricted. Both countries have property systems that protect private property rights; they just define those rights differently. Similarly, in some countries, the means of production (factories, natural resources, etc.) are controlled by the government but private individuals can buy and sometimes sell consumer goods. These are typically called socialist countries — but even socialist countries do protect some private property rights.

For our discussion, let's define a system of private property as one in which most material resources (and often other things like ideas or inventions) are divided up, with each one belonging to an individual who has authority to decide how it will be used, subject to general laws protecting community welfare. This describes the United States' property system pretty well. Importantly, even in the United States, property rights are not and cannot be unlimited. We are never free to use our property to create a health hazard or to commit a crime, for example. But in a system that recognizes private property rights in land, private individuals — not a social group or the government — make most of the decisions that affect the land and other things they own.

Why is such a system ethically justified? Obviously, one argument for a system of private property is that it facilitates the use of markets, and markets are very effective ways to distribute goods efficiently. Even socialist countries often create markets in consumer goods because they are superior to a system of government distribution for getting goods to the people who most want them.

Still, even their greatest admirers admit that markets are not always efficient (their environmental drawbacks are one of the main topics in the field of environmental economics.) And from an ethical point of view, this system of private property is vulnerable to the charge that it allows one to hoard one's wealth rather than give it to people who really need it. Why would we set up a property system that allows such selfishness? (Note that economists, who usually assume that economic efficiency is an important value, tend to shy away from the related value question: whether seeking to increase your own wealth is ethically justified. They don't see this ethical question as part of their discipline.)

The most common ethical argument for a private property system refers back to the concept of rights. Recall our discussion of rights in Sect. 3.3: Under liberal social contract theory, we protect rights in order to ensure that humans can exercise rational autonomy, the freedom to govern their lives as they choose. Obviously, part of governing your life involves creating a home, getting food to eat, clothes to wear, and other necessities. More generally, for most people, an important part of a good life is planning and carrying out projects that involve manipulating the physical world or creating ideas. Protecting private property rights is one way to make sure people have the freedom to carry out such projects. Yes, some people may use that freedom to make ethically poor judgments. They may refuse charity to people who really need it, or spend their wealth on frivolous activities. But respecting humans' rational autonomy means respecting their freedom to make poor choices.

We can find even stronger philosophical arguments for private property in the works of the German philosopher G.W.F. Hegel (1770–1831). Hegel was not a liberal philosopher; he's part of the German idealist tradition. But his theory of property has been extremely influential because it addresses an important psychological dimension of ownership. Hegel argues that the institution of private property is essential to human freedom because individuals express personality and become self-aware largely through controlling things. More specifically, humans give purpose and meaning to things by taking control of them, by embodying human intentions in them. Through this activity, we make the nonhuman world less alien and more human; we make it a place where we can feel at home. Even more importantly, we express ourselves through this taking of possession; we make our will concrete and public so that others can recognize who we are and what we value. This process of taking control of things is therefore a key avenue to developing our personality. It's also vital to creating a meaningful connection to the nonhuman world — which is why protecting property rights might be important to a community dedicated to environmental stewardship.

Hegel argues that the development of personality through private ownership is an essential step on the path to a fully ethical community. That's a much longer story, but this brief account of his theory is sufficient to explain its appeal. It tells us why we might develop strong psychological attachments to the things we control even in the absence of legal property rights, and why we can develop a deep connection the land we live on.

Because property rights are so important to freedom and to our sense of self, some people consider private property rights to be natural rights — that is, they are part of the natural freedom that government was invented to protect. This is the traditional liberal view of property, and most liberal governments give fairly strong protections to private property rights. Property rights are an important part of the "common good" that the government is supposed to protect. Others argue that property rights are not natural rights at all; they are simply legal devices created by the government to serve the common welfare. Under this view, private property rights are justified only to the extent they serve the common good — and the common good is usually defined in terms of peoples' ability to meet their needs, not by their degree of freedom from government control. Under either view, however, property rights are clearly useful. They help to protect individual freedom and they can be part of an efficient economic system.

6.2.2 Property Rights and Freedom in the United States

Are property rights really important to freedom? The history of the United States suggests that they can be. Recall Alice Walker's words (Sect. 4.2.1): "The history of all black Southerners is a history of dispossession. We loved the land and worked the land, but we never owned it." While that's a bit of an overstatement, it's true that a major element of slavery and Jim Crow was ensuring that African Americans had fewer and less secure property rights than white Americans did. Property is power; oppression requires that the oppressed be denied the opportunity to acquire the economic, political, and social power that goes with land ownership. African Americans were not the only ones who found land and freedom connected. Native Americans were also displaced and dispossessed of their ancestral lands, with devastating consequences for their communities and political status. Most of the Mexicans living in what became the American Southwest lost their land when that territory was annexed under the Treaty of Guadalupe-Hidalgo in 1848. Many Japanese-Americans lost their land during World War II, when they were forced into internment camps. Because of this history, struggles for equality and freedom in the United States usually involve efforts to take possession of the land and defend property rights. This is why property rights are generally included among the civil rights that the American constitutional tradition protects.

6.2.3 Property Rights and Environmental Protection

Despite their importance to freedom, even the most committed defender of property rights would admit that the government can put some limits on them. Recall that under social contract theory, liberal governments are limited to protecting human autonomy. That includes protecting property rights. For example, the government should prevent a factory from polluting the water that flows into your stream or the air you breathe, if that water and air is part of your property. Many environmental laws can be justified under social contract theory as protections for property rights. Such laws prevent people from *externalizing* the environmental costs of economic activity (that is, imposing their pollution on other people).

But environmental laws can also serve to redistribute wealth in order to achieve greater social equality. Liberal governments can justify redistributing property from the rich to the poor, up to a point: Protecting individual autonomy can include making sure we all have the things we need to live an autonomous, choiceworthy life. After all, there's little point in having a right to property if you don't have the money to buy anything. This is how governments justify taxing people in order to provide a basic social safety net — a welfare system — for the poor. The same rationale can justify public control of some natural resources. For example, the government establishes national parks so that every citizen (and not only the very wealthy) can enjoy natural wonders like the Grand Canyon and Yellowstone National Park. (Recall Elizabeth Anderson's argument that natural wonders like these should not be treated as commodities to be bought and sold. Government control can ensure that they are treated appropriately, as part of our collective national heritage.)

The government may also control resources in order to serve the needs of future generations. Markets are pretty good at serving the needs of the present generation, but they don't have any way to take into account interests of people who don't exist yet. Governments may therefore take control of some resources (like forests or waters) and limit what people can do with their private property, in order to ensure that they will be available for future citizens. (Obviously, this is justified only if we believe we have duties to future generations, as discussed in Chap. 5).

Finally, the government might enact environmental regulations in order to protect the interests of foreign nations. For example, the United States has entered into treaties to protect migratory birds, fish and marine mammals, and other animals that cross international lines. These treaties help the U.S. remain on good terms with those nations. So restricting property rights in the U.S. might be part of a more comprehensive foreign policy.

Of course, any of these reasons for protecting the environment can be challenged as illegitimate. Some people, including many Americans, doubt that the government should be redistributing wealth from rich to poor, or from the present to the future. They question whether natural wonders are better off in the hands of government. These people do not necessarily disagree that protecting the environment is important. One may believe strongly that individuals have an ethical obligation to be good stewards of the earth but also hold that government should not force us to meet that obligation. Perhaps one's duty to the earth, like one's duty to God, is a private matter that the government shouldn't interfere with.

This point brings us back to Judy K.'s dilemma about the sand mining. Judy may be persuaded by Maria that the sand mining operation is not good stewardship. Perhaps it will damage the ecosystem, or simply shows a lack of appreciation for the natural beauty of this special piece of mid-Michigan. But she may also be reluctant to ask the government to deny the owner a permit to do the mining.

She respects the owner's freedom, including his freedom to make bad decisions. She is also a landowner, and she values the freedom to make decisions about land management. She has come to appreciate how making her own decisions — even bad ones — and taking responsibility for the consequences has strengthened her connection to the land and to her community. Taking responsibility for her decisions leads her to explore new questions and learn more about the ecosystem she and her neighbors depend on. In short, she values environmental decision making as part of a good life, and heavy-handed government intervention could take those decisions out of her hands.

Fortunately, heavy-handed government intervention isn't the only way to encourage people to be better stewards. Persuasion might be just as effective. What sort of arguments might Judy make to persuade the owner to change or abandon his plan?

6.2.4 Community Interests in Private Property

One approach is to talk to the owner about property rights. The owner of the sand-mining operation clearly has rights in his property that he expects others to respect. But to be consistent, he also has to respect the rights of others, and other people in the community may have rights in his property. For example, state law may require property owners to give people access to lakes and rivers on their property, or to provide a public right-of-way over it. Even where the law is silent, some communities have long-standing customs concerning how the public can use private property. In some places, it's acceptable to enter private property to pick wild fruit, hunt game, or go fishing. Landowners must learn to respect that custom and avoid interfering with such public rights.

A special relationship to the property owner can also create an interest in private property. For example, Judy believes that her tenants and her children have interests in her property, and she tries to protect those interests. Moreover, if she sees herself as part of a biotic community, she may conclude that nonhuman members of that community have rights that she is bound to respect. These rights may not be enforceable in court, but they may carry moral weight with the property owner.

Once we start listing all the legal or customary restrictions on how one can use one's property, and all the people or nonhuman beings who can have an interest in that property, the concept of ownership starts to look quite complicated. Clearly owning a piece of property doesn't give you compete sovereignty over it. It gives you a set of privileges but also a set of responsibilities. Moreover, that collection of privileges and responsibilities may be different for each different piece of property.

Some scholars have developed theories of property that takes these complications into account. For example, political theorist Jeremy Waldron notes that things capable of being owned vary dramatically, and it's unlikely that the same concept of ownership could apply to all of them. It makes sense, for example, to treat air and water differently than rocks and sand, and to have special rules for large-scale natural resources, like oceans, the atmosphere, and Antarctica. Legal scholar Eric Freyfogle agrees, arguing that we should discard the idea that property rights can be defined without reference to the thing being owned. Land parcels, for example, differ greatly, so private landowners have different rights and responsibilities depending on whether they own beachfronts, wetlands, or critical wildlife habitat. These theories are attempts to transform legal and ethical theory, to integrate it with the science of ecology. So in the future, we might develop a system of property rights rooted more in ecology and environmental values.

For now, it's important to understand that owning a piece of property doesn't give you the unlimited right to do anything you want with it. A proper regard for the community's interest in privately owned land requires a responsible property owner to consider how his plans will affect others. The owner of the sand-mining operation might respond to such arguments, since presumably he wants others to respect his interests as well.

6.3 Stewardship and the Good Life

Even if the property owner doesn't respond to arguments about the rights of others, he might be open to arguments about how being a good steward can enrich his own life. After all, people buy property with the understanding and hope that they will become responsible for it; carrying out such responsibilities is, for many people, an important part of the good life. Taking care of something can be a valuable basis of self-respect and can earn you the respect of others.

Stewardship is one of the social ideals that Elizabeth Anderson talks about, like "animal friendship" (Sect. 4.2.4) It's realized by engaging in social practices that express this ideal, like tending your garden, mowing your lawn, observing catch limits for your lake, or following "leave no trace" principles when you go hiking. Just as taking care of your dog is a way to realize the ideal of animal friendship, so taking care of a place is a way to show that you value that place appropriately.

The ideal of stewardship has been part of the Western ethical tradition for thousands of years. The concept is central to the agrarian tradition, an ethical tradition dating back at least to the Roman civilization and including such classic works as Cato's *De Agricultura* and Virgil's *Georgics*. Agrarian philosophers celebrate the virtues and rewards of farming as a way of life. They argue that a life spent tending the earth cultivates desirable character traits (like diligence, humility, and self-reliance) and offers a host of healthy, fulfilling pleasures. Contemporary advocates of the stewardship ideal extend the concept beyond farming; stewardship includes tending to the wilderness and creating habitable cities where humans and nonhumans can thrive.

The Christian tradition also celebrates the idea of stewardship. Christian theologians have characterized the earth as God's creation, a gift that demands careful and reverent care. Tending the earth is for many Christians a sacred duty. For example, the Evangelical Environmental Network website describes their organization as "grounded in the Bible's teaching on the responsibility of God's people to "tend the garden" and in a desire to be faithful to Jesus Christ and to follow Him" ("Our Witness," Evangelical Environmental Network website 2018). Similarly, in 2015, Pope Francis issued the encyclical letter "Laudato Si," which called for Catholics to care for the natural environment and repair ecological damage. Other faith traditions that characterize the Earth as the creation of a god or gods typically value stewardship as well.

With such a rich and impressive pedigree, it's not surprising that the ideal of stewardship resonates with a wide range of people. But philosophers have only recently begun exploring the ethical foundations of this ideal. One important line of philosophical investigation is called "the ethics of care." Feminist philosophers have led this inquiry into ethical principles rooted in relationships of care, such as the relationship between mother and child. Carol Gilligan's groundbreaking book, *In A Different Voice*, argued that traditional ethical theories assume that ethical reasoning requires one to distance oneself from personal relationships and adopt a more objective point of view. But she identified an alternative mode of ethical reasoning that begins with careful, loving attention to the people one is responsible for. This ethic emphasizes sensitive and emotional response to particular needs rather than impersonal respect for universal rights.

Feminist environmentalists (sometimes called ecofeminists) such as Marti Kheel have applied the ethics of care to the natural world. The idea of developing a relationship of care with a particular place and responding to its needs corresponds quite well with the ideal of stewardship. As philosopher Bruce Jennings puts it,

> Care begins with the particulars of society, culture, and psychology—its starting point is the latent possibilities of *a given place at a given time* and with ongoing forms of meaningful agency… It builds on senses of historical memory and tradition, and it feeds on the gratitude felt when one recognizes the service and contributions that others have made to one's way of life. Care begins with the recognition of symbiotic interdependence and then intervenes in—interrupts—an ongoing form of life in order to be present to the need, vulnerability, and suffering it contains, finally winning through to a better kind of species activity and life well-lived (Jennings 2017, p. 11).

The stewardship ideal does have some critics, of course. Some environmentalists see it as focused too much on human needs. Farmers, after all, tend the earth in order to feed humans, not for the earth's own sake. Environmentalists and philosophers both worry that focusing on the needs of a particular place may prevent us from attending to large-scale issues, thus elevating local concerns above more general, common interests. However, precisely because it's rooted in human needs, local concerns, and an emotional connection to the land, the stewardship ideal does seem well-suited to motivate action to protect the environment.

The Agrarian Tradition in the United States

Agrarianism is a long-standing theme in Western thought, centering on the belief that man's natural calling is to cultivate the earth. Christian agrarians have traditionally cited Genesis 1:28 to support the claim that God gave humans stewardship over nature, including the duty to domesticate animals and cultivate plants; some would characterize humans as "co-creators" whose function is to "finish" Creation, transforming the wilderness into a garden suitable for human flourishing.

By the eighteenth century, agrarianism permeated American religious and political discourse. This tradition has a dark side. For example, colonists often tried to justify displacing the natives on the grounds that the earth rightfully belongs to those who use it as God intends (in their view, for European-style agriculture). But it also served as a democratic ideal. Thomas Jefferson claimed that farming creates a character ideally suited to republican government, so our policies should favor small, family farms. Owning productive land provides citizens with economic and therefore political independence. Agricultural labor, under this view, cultivates virtues conducive to good citizenship, including self-sufficiency, industriousness, humility, spirituality, and prudence.

Contemporary environmentalist and social critic Wendell Berry draws on agrarian themes in his eloquent defense stewardship and a place-based ethics. Berry envisions the good life as a life rooted in place, where responsibility to particular things and people can be enacted in a meaningful way. Living in place by relying on local foods, markets, and services allows us to develop our consciousness of our interdependencies and should enhance our sense of responsibility, thus developing our environmental virtues.

Berry's agrarianism centers on a conception of the human condition as one of interdependence and vulnerability. That understanding grounds his vision of the good life: a life of stewardship in community with co-equal citizens and oriented toward preserving the environmental and social conditions necessary for us to lead fully human lives.

6.4 The Problem of Sand Mining Revisited

Judy never got the chance to have a one-on-one conversation with owner of the sand mining company about his proposal. Instead, her neighbors used the democratic process to conduct a more formal, but more inclusive, conversation. After several public hearings and a lot of email conversations, they persuaded the township board to pass on ordinance regulating mining. Mining is not prohibited, but to receive a permit, owners have to do a number of things to protect the ecosystem. The ordinance establishes setbacks and buffers to protect waterways, regulates the noise levels, and requires the company to restore the area to its previous condition when the mining is completed.

Restricting (or defining) property rights through regulation is one way to ensure that the community's interest in the land is protected while giving the property owner a chance to use the property as he wants. But it does pose complications, because an ordinance doesn't enforce itself. Protection of those community interests will require vigilance from the community and willing cooperation from the landowner as long as the mining is going on. This operation will create a long-term relationship between the community and the mining company.

There is another way to use property rights to achieve environmental goals, though: Some of the neighbors might buy the land themselves, so they won't have to deal with this company at all. They can assert their property right to preserve the land as it is. That may sound like an appealing strategy, but it does have a cost. Buying the land represents a refusal to create an on-going, cooperative stewardship relationship with the sand-mining company. Local environmental values will be protected, but is this the best outcome for the larger community? After all, the company owner may simply move on to another location, where the citizens are less environmentally conscious. And the Spring Lakes community will lose an opportunity to learn from and influence values of the owners of the sand-mining company.

One point to take away from this discussion is that taking care of any place on earth is never just an individual task; it involves the whole community in some fashion. There are *and should be* many different interests and values involved. Therefore, environmental stewardship involves you in complex and often demanding relationships — not only with the natural world but with other people. It requires you to develop virtues and social skills, to learn how to work with others toward a common goal. This is precisely why so many people see stewardship as an important part, and perhaps even the foundation, of a good, choiceworthy life.

Further Reading

Berry, W.: The Unsettling of America. Sierra Club Books, San Francisco (1977)
Evangelical Environmental Network.: http://www.creationcare.org/witness. Accessed January 25, 2018, DOI: https://doi.org/10.1055/s-0038-1632381
Francis.: *Laudato Si.* May 24, 2015. http://w2.vatican.va/content/francesco/en/encyclicals/documents/papa-francesco_20150524_enciclica-laudato-si.html. Accessed January 28, 2018
Freyfogle, E.: Agrarianism and the Good Society. University of Kentucky Press, Lexington (2007)
Jennings, B.: Ecological care. Minding Nature. **10**(2), 4–11 (2017)
Kheel, M.: Nature Ethics: An Ecofeminist Perspective. Rowman & Littlefield, New York (2008)
Waldron, J.: The Right to Private Property. Clarendon Press, Oxford (1984)
Worrell, R., Appleby, M.: Stewardship of natural resources: definition, ethical and practical aspects. Journal of Agricultural and Environmental Ethics. **12**, 263–277 (2000). https://doi.org/10.1023/A:1009534214698

Chapter 7
Valuing Landscapes

Contents

7.1 The Problem of the Diseased Oaks

Judy K. has a dilemma. The black oaks on the northwest side of the property are dying, probably because of a fungal infection or oak wilt. She has consulted a forester, who advised her to remove the damaged oaks. Not only would removing the trees be good for the forest's health, she could sell the lumber for several thousand dollars. Unfortunately, this operation would probably disturb one of the more interesting features of the place: there are two piles of rocks in the forest that might be burial mounds left by the Potawatomi people who once lived here. The rocks aren't visually impressive and might just be rock piles left by the previous owner of the property. But if they are burial mounds, Judy wants to respect their historical and cultural significance. And the tree removal would also leave some unsightly scars, trample the underbrush, and generally disrupt the green peacefulness of the forest. These impacts would be temporary, of course, but they would annoy some of her tenants. In fact, a number of the tenants are opposed to any active management of the forest; they prefer that it be left in what they consider its "natural" state. They object also to clearing the undergrowth from the forest and putting up fences to keep the geese off the lawns near the pond. There have been similar controversies over what to do with the marsh: Some tenants would be happy for it to be drained (although state laws protecting wetlands would probably prohibit that). Others want it left alone. One tenant has proposed turning it into a small nature education resource, with walkways and interpretive signs.

K. K. Smith, *Exploring Environmental Ethics*,
AESS Interdisciplinary Environmental Studies and Sciences Series,
https://doi.org/10.1007/978-3-319-77395-7_7

All of these management decisions require Judy to weigh ecological, aesthetic, scientific, historical, economic, and other kinds of value the landscape holds for her and her small community. Should she put ecological health over the historical and recreational value of the forest? If so, what constitutes a "healthy" ecosystem: one that is actively managed for biodiversity or one that is left to go its own way without interference? How should aesthetic and historical values be weighed against ecological and economic values? How should she resolve questions like this?

7.2 Landscape and Meaning

Landscapes carry many different kinds of instrumental value. That is, humans (and other beings) can gain many kinds of goods from a landscape, including economic, aesthetic, spiritual, historic, cultural, and ecological benefits. A river, for example, might be considered a sacred site, a recreational resource, a part of the transportation system, and a critical ecosystem. These meanings can all contribute to a *sense of place*, which some theorists consider an important foundation for environmental ethics. Having a strong sense of place can bring meaning to your life and motivate action on behalf of the natural environment. It's a distinctive good that can be achieved through one's relationship with the natural world.

Arguably, it is part of a land manager's job to create and maintain a sense of place. But any given landscape can acquire many different kinds of meanings and value. Which meanings and values should be prioritized? Should Judy manage Spring Lakes to achieve aesthetic values, ecological values, economic values, or cultural and historical values?

Unfortunately, we can't say that only one of these values (like ecological value) is real and the others are just a matter of opinion. True, the ecological significance of a river or forest can be verified scientifically, which makes it seem objectively true. The economic value of a natural phenomenon also seems to be quantifiable and therefore objective. But we can verify other kinds of value objectively as well. For example, the spiritual significance of a natural object can be verified through theological or anthropological investigation. Such investigations show that the Ganges River in India is sacred to Hindus and the Devil's Tower in Wyoming is sacred to several tribes of the Plains Indians. Those facts are as objectively true as the facts about their ecological importance.

Of course, other social groups may see different meaning in those landscapes. This is simply a curious feature of our social world: The same object can carry many different meanings, all of which are compelling and well-justified. These differing interpretations of the landscape are the source of much conflict in environmental policy.

This chapter will focus on two persistent and important sources of interpretive conflict: the meaning and value of *wilderness* and how to judge the aesthetic value of a landscape. But before we venture into those debates, we must address a more basic question: Why is the meaning of the landscape an ethical question at all?

Normally we don't think someone who prefers an Adam Sandler movie to *Citizen Kane* has made an ethical error (although we might question his taste). Nor would we consider a poor, misguided interpretation of *Citizen Kane* to be an ethical failure. The interpretation of movies doesn't seem like a matter of ethics. So how does the interpretation of the landscape become a matter of ethics?

To answer that question, recall the broad sense of "ethical" that we established at the beginning of this text: Ethics concerns how to live a good, choiceworthy life. How one interprets the landscape may be relevant to that question in three ways.

First, a choiceworthy life in relation to the natural world surely includes the capacity to appreciate nature. That is, all other things being equal, your life is better and happier if you can take pleasure from a beautiful sunset, find meaning in a forest glen, or feel wonder at a waterfall. Of course, nature is not the sole source of such pleasures, but it is one important source. Without this capacity to appreciate the complex meanings and values of the landscape — a sense of place — one's life is diminished to some extent. If that's true, then a good life would include learning to find many kinds of meaning and value in the landscape. If one wants to live well, one should learn to recognize and appreciate the land's historical, aesthetic, and other kinds of values. (Whether appreciating the landscape properly requires you to study ecology is a question taken up below.)

Second, even if you don't value a landscape very much, other people do. We've already noted that many different people may have an interest in any given piece of land. The good land manager would recognize these other interests and respect the values that others find in the land. For example, Judy K. is reluctant to disturb the burial mounds on her property. She doesn't share the cultural or religious beliefs that gave these graves spiritual significance to the people who made them. But she does want to treat other people's religious beliefs with respect. She also recognizes that these mounds have historical importance to many people, even though she herself may not be very interested in this history. Respecting other persons' meanings is part of respecting other persons, which is surely an ethical obligation.

Third, and more controversially, one could argue that failing to appreciate the meaning and value of the landscape is a kind of *moral blindness*. I take this term from Lawrence Blum, who in turn borrows the idea from philosopher Iris Murdoch. But the concept of moral blindness is actually much older than that. The early American abolitionists argued that people born into a slave society often suffered from this kind of blindness, an inability to recognize or fully grasp the moral ugliness of slavery. People who could watch a slave being beaten without showing any repugnance or horror deserved moral censure. Such people showed a moral insensitivity that seems objectionable in itself. One wouldn't want to be so insensitive to injustice, even if one were powerless to stop it. Blum offers a more contemporary example of a person who fails to recognize a blatant example of racism; such a person seems culpably blind to morally important features of his social world.

Aldo Leopold, Henry David Thoreau, John Muir and many other nature writers have similarly condemned blindness toward nature's beauty and meaning. Consider this passage from Aldo Leopold's essay "Illinois Bus Ride":

> I am sitting in a 60-mile-an-hour bus sailing over a highway originally laid out for horse and buggy. The ribbon of concrete has been widened and widened until the field fences threaten to topple into the road cuts. In the narrow thread of sod between the shaved banks and the toppling fences grow the relics of what once was Illinois: the prairie.
>
> No one in the bus sees these relics. A worried farmer, his fertilizer bill projecting from his shirt pocket, looks blankly at the lupines, lespedezas, or Baptisias that originally pumped nitrogen out of the prairie air and into his black loamy acres. He does not distinguish them from the parvenu quack-grass in which they grow. (Leopold 1966, p. 125)

Of course, failing to appreciate wildflowers or an ancient forest is hardly on the same scale as failing to appreciate the horror of human slavery or racism. But it may be a much smaller version of the same kind of insensitivity: An inability to recognize phenomena of moral importance. If you believe you have a duty to nonhuman beings or an obligation to protect valuable ecosystems, then the failure to notice harms to these entities is, perhaps, a moral failing. Philosopher Yuriko Saito offers a similar argument: Learning to appreciate nature in its own terms requires us to develop an important moral capacity, the capacity to overcome the confines of our own perspective — to recognize and understand another being's reality through sympathetic identification. So failure to appreciate nature may be evidence of a moral defect. Or, to use Elizabeth Anderson's terms, the inability to appreciate nonhuman nature constitutes a failure to value the nonhuman world appropriately, which is just a moral mistake in its own right.

If this argument is persuasive, then how we value the landscape is relevant to ethics. Therefore, disciplines that teach one how to see and appreciate the natural world — the arts and literature in particular — are important to ethical development (and especially environmental ethical development). This is one reason that courses on nature writing and field drawing are often integral parts of an environmental studies curriculum.

But as anyone making land management decisions will discover, how we see and value the landscape is also a source of much of the conflict over environmental policy. Indeed, this may be the most contentious, complex, and challenging areas of environmental ethics.

7.3 The Value of Wilderness

One of the persistent debates in environmental policy centers on the value and meaning of "wilderness." Americans have traditionally placed special value on wild landscapes. The American environmental movement began, in part, as a wilderness preservation movement. Advocates for wilderness like John Muir and Aldo Leopold have helped us to appreciate the aesthetic, moral, and spiritual value of wild places. From Huck Finn's fictional adventures on the Mississippi to the modern wilderness journey recounted by Cheryl Strayed in her 2012 book *Wild*, Americans have learned that the wild is a place to test one's self-reliance, seek spiritual insight, and discover one's true self.

But what exactly is "wilderness"? The American Wilderness Act, enacted in 1964, defined it this way: "A wilderness, in contrast with those areas where man and his own works dominate the landscape, is hereby recognized as an area where the earth and community of life are untrammeled by man, where man himself is a visitor who does not remain." This legal definition probably captures the common understanding of wilderness pretty well. But it rests on a problematic contrast between what is natural and what is human. Dividing the world this way obscures the fact that humans, too, are the product of nature, and very little of the natural world can reasonably be considered free of human influence.

In fact, the debate about wilderness is part of a deeper controversy in environmental ethics: Does a phenomenon have greater value because it is natural? One reason it's hard to answer that question is because "natural" can have so many different meanings. Andrew Brennan and Y.S. Lo have helpfully offered the following four definitions:

N1: subject to causal laws of physic, chemistry, etc.
N2: spontaneously self-originating and self-maintaining according to principles of evolution and ecology (not dependent on human intervention).
N3: In keeping with one's species-specific nature.
N4: Not influenced by human reason or reflection. (Brennan and Lo 2010, pp. 118–120.)

You see the problem: A golf course is natural in the first sense but not in the second or fourth. A cow is natural in the first and third sense, but not in the second or fourth. Wilderness is natural in the first and second sense, but it's not clear that there really are any wild places left that don't depend on human intervention and rational planning to remain in their condition.

The second reason it's hard to know how to value "natural" things is ambiguity over what kind of value we're talking about. Some environmental ethicists insist that being natural confers moral value. That is, anything that comes from nature is morally good. But that position faces a difficulty explained by the English philosopher John Stuart Mill. In his 1874 essay *On Nature*, he argues that "natural" should not be considered synonymous with morally good, because many actions or conditions that are natural in the above senses are bad — undesirable or morally disvalued. After all, it may be natural to distrust strangers or become violent when angry, but that doesn't mean such reactions are always morally justified.

On the other hand, it is much easier to argue that being natural in the above senses confers aesthetic value. We discussed aesthetic value briefly in Sect. 4.2.5, noting that aesthetic value may depend not only on the physical characteristics of a phenomenon but also on how it came into being, its history and place in a larger context. Just as a reproduction has less aesthetic value than an original artwork, so might a landscape produced by human artifice have less value than a landscape produced entirely by natural forces.

In addition to all the ambiguities around the concept of "natural," a number of contemporary environmentalists have questioned whether the wilderness as traditionally defined is really what environmentalists should be focusing on.

William Cronon shocked the environmental movement in his 1995 essay "The Trouble with Wilderness," arguing that "wilderness" is a socially constructed concept that reflects the social position of a narrow demographic: wealthy, white, male Americans — that is, the people who had the means to spend their leisure time camping, hunting and fishing, as opposed to the people who lived on and worked the land. Cronon worried that the concept of wilderness may teach us to be too dismissive of more humble, ordinary landscapes, which are also rich in aesthetic, historical, and ecological value. Bill McKibben further questioned the value of the concept in his 1997 book, *The End of Nature*. McKibben pointed out that anthropogenic climate change has affected every ecosystem on earth, so if "wild" means pristine and "untrammeled by man," then there are no wild places left to protect.

Moreover, the common conception of wilderness as "untrammeled by man" seems to leave no possibility of wilderness restoration, the practice of returning a disrupted landscape to its former conditions. For example, consider the Don Edwards San Francisco Bay National Wildlife Refuge. Located on the San Francisco Bay near Fremont, the Refuge spans 30,000 acres of open bay, salt pond, salt marsh, mudflat, upland and vernal pool habitats. Once heavily mined for salt, the refuge is in the process of being "restored." Although this flat, muddy expanse hardly resembles a traditional wilderness, the Refuge has considerable ecological value. It hosts over 280 species of bids each year and provides critical habitat to resident species like the endangered California clapper rail and salt marsh harvest mouse. Due to its convenient location in the Bay Area, hundreds of thousands of people visit the Refuge each year.

Some philosophers (notably Robert Elliott and Eric Katz) insist that a restored landscape like the Don Edwards Refuge is no more wild than a golf course. Humans can create only a fake (and very much inferior) version of nature's work. On the other hand, defenders of restoration argue that the real value of wilderness lies not in its freedom from human management but in its specific ecological services or aesthetic qualities, and these can (sometimes) be recreated or enhanced through human management. But that defense assumes that the quality of being "pristine" is not what we value about wilderness.

7.4 The Value of Biodiversity

Perhaps the environmental movement needs a different goal. Even as the debate over the meaning of wilderness has been going on, the environmental movement has quietly been shifting its focus from wilderness to biodiversity. Biodiversity (like "ecosystem" and "species") is a term borrowed from the sciences, and as such it seems to be less controversial than the concept of wilderness. But when "biodiversity" moves out of the scientific arena and becomes a public policy goal, it becomes as difficult to define as "species" and "ecosystem."

Ecologists may define *global biodiversity* as the variety of different organisms in the world, or they might include also the variety of different genotypes and ecosystems.

Most people would agree that preserving biodiversity in any of these senses is generally a good thing to do. As we learned in our earlier discussion of species preservation, one rationale for preserving biodiversity is that we want to live in a world that is rich in many different kinds of goods and experiences. Protecting different ecosystems and organisms can contribute to that goal. But when we turn from theory to practice, it turns out that what we're aiming for when we preserve biodiversity is not merely the total number of different organisms. We are also concerned about which organisms *belong* in which ecosystem, in their functional relationships with each other and with the land. We may even exclude some organisms altogether, like genetically engineered lab mice, because they aren't "natural." So we are faced with questions like: Should we irrigate the Arizona desert in order to create fields, forests, and gardens? Is the Brazilian rainforest more valuable than a desert because it supports more species — or does the desert have aesthetic and ecological values comparable to a rainforest? If biodiversity simply refers to the number of different species in an ecosystem, perhaps we should introduce exotic species into ecosystems that lack species richness. Or perhaps we should develop new organisms in order to increase genetic richness. On the other hand, perhaps we value having a variety of different ecosystems rather than a greater number of species.

Of course, the fact that "wilderness" and "biodiversity" are hard to define doesn't mean we should abandon attempts to preserve them. We just have to clarify exactly what kinds of wild landscapes and ecological diversity we favor, in what context, and why. Those are ultimately value judgments, and therefore call for ethical (not just scientific) thinking. Clarifying these judgments will help us to justify our efforts to preserve landscapes and figure out how to balance biodiversity against other important values, like aesthetics.

7.5 Aesthetics and Ecology

We learned in Sect. 4.2.5 that aesthetic qualities are qualities of an object or experience that contribute to (or detract from) one's appreciation of it. Not surprisingly, the aesthetic value of landscapes is another area of disagreement and conflict. Some people prefer a highly managed, park-like landscape; others prefer to let nature run wild. And who can say what the next generation will prefer?

Philosopher Martin Krieger explored the meaning of wilderness, rarity, and naturalness in his well-known 1973 article, "What's Wrong With Plastic Trees?" The essay responds to a suggestion by some highway planners that it would be cheaper to install fake trees along a highway meridian than to maintain real trees. Krieger wondered, "Why not?" Creating artificial landscapes to give people the aesthetic experience of nature might, in some cases, be the best way to balance aesthetic and economic values. His argument suggests that "naturalness" or "pristineness" are aesthetic values that we cannot always afford.

The idea of creating artificial versions of natural phenomena sounds like a futuristic nightmare to some people. But Krieger points out that many of our iconic so-called wild landscapes, like Niagara Falls, are carefully engineered. Does knowing that such a landscape is actually highly managed detract from its aesthetic value? Is a "fake" landscape comparable to a forged painting — or do we need to use different aesthetic criteria for natural phenomena?

These are questions on which people may easily disagree. Aesthetic judgments like this often seem highly subjective and perhaps irreconcilable, and therefore a poor basis for public policy. Of course, we don't want to overstate the problem: Just as we can make reasonable, well-supported arguments about the quality of literature or art, we can also analyze and evaluate the aesthetic value of a landscape. Aesthetics is an area in which we can seek if not objective truth at least intersubjective consensus. (For example, we may not agree on whether plastic trees in general are a good idea, but we may be able to agree that some plastic trees are downright ugly). Nevertheless, a policy maker may prefer to rest land management decisions on something more stable and objective than a possibly temporary consensus.

That search for an objective, stable element in aesthetic judgment has led ethicist Baird Callicott to argue that ecological health is the foundation of the aesthetic appeal of a landscape. That is, he claims that we should see an ecologically healthy landscape as more beautiful than an ecologically damaged landscape. Learning about ecology, under this view, is essential to developing a proper nature aesthetic. Just as an art critic must learn to appreciate the more complex aesthetic values in a work of art — its composition, color palette, technique, historical allusion, and so on — so the nature lover must learn to appreciate the evolutionary history and ecological function of natural phenomena. "Ecology, history, paleontology, and geology each penetrate the surface of direct sensory experience and supply substance to scenery," he writes in his essay "Leopold's Land Aesthetic." A proper aesthetic appreciation of nature, for Callicott, requires *biological literacy* (Callicott 1989, p. 241.) This argument supports the view that a natural landscape has more aesthetic value than one produced by human artifice. Environmental aesthetics therefore differs from how we judge works of art, where human artifice is precisely what we are evaluating.

Under Callicott's view, an ecologically healthy landscape would have greater aesthetic value than a degraded one. An old-growth forest would be far more appealing than a picture-perfect golf course, and a yard filled with native grasses would have to be considered more pleasing than the traditional well-manicured lawn. Such a revolution in aesthetic values might bring the ecological and aesthetic value of the landscape together more consistently. But Callicott's view isn't universally endorsed by environmental philosophers. Stan Godlovitch, for example, makes a powerful argument against basing one's nature aesthetic on scientific analysis. He contends that the real aesthetic value of nature is in appreciating its mystery. Encountering the ineffable and unknowable in nature can remind us of our limits and proper place in the grand scheme of the cosmos. Godlovitch's argument might lead us to conclude that we shouldn't learn too much about evolutionary biology and ecological science, precisely because such scientific knowledge might ruin the mystery (in the same way that learning too much about literary technique may make it difficult to enjoy a good story.) Nevertheless, Godlovitch agrees with Callicott that the task of environmental aesthetics is to appreciate the works of nature, not the accomplishments of humans.

Even if we could bring aesthetics and ecological value into perfect alignment in this way, however, we would not eliminate conflict over the meaning of the land. Social groups would still fight over which values should predominate in land management decisions. Resolving these disputes is not just a matter of philosophical analysis; they often implicate social justice.

7.6 Landscapes and Injustice

Land managers soon learn that landscapes are sites of social conflict, and powerful social groups are more likely to win these fights. As a result, the landscape generally reflects the meaning and values of the dominant group in society. As the previous chapter pointed out (Sect. 6.2.2), minority groups in the United States have often been denied secure property rights. Part of that dispossession is the tendency to erase those groups' meanings from the landscape. William Cronon makes this point in his essay on the meaning of wilderness: To Americans of European descent, places like Yosemite and Yellowstone may look like pristine wilderness — places "untrammeled by man." But to the Native Americans who lived there, those places weren't untrammeled at all. They were the ancestral homes, sacred landscapes rich in cultural and historical meaning. The creation of national parks in Yosemite and Yellowstone involved erasing those meanings and redescribing those places as empty of human inhabitants. It's hardly surprising to learn that the actual human habitants were also removed from these places.

We suggested above that making decisions about which landscape values to protect and promote had something to do with respecting persons, which suggests that it is a matter of social justice. Judy K. wants to respect any burial mounds that might be on her property because she feels a sense of obligation to respect the group (the Potawatomi people) that created them. But what are our obligations to respect or preserve the meanings of other groups? In what sense does failing to respect the meaning of a place constitute a "harm"?

First, it's important to remember that conflicts over the meaning of the land do not need to be zero-sum. That is, it is sometimes possible to respect *all* the values of the land — even the conflicting interpretations of different social groups. For example, Devil's Tower, mentioned above, is a national monument managed by the National Park Service. It is also sacred to several Native American Plains tribes, including the Lakota Sioux, Cheyenne, Crow, and Kiowa. The Tower is a popular site for rock-climbers, but a number of Native American leaders consider climbing the Tower to be a desecration. Many of the climbers, in contrast, insist that rock climbing can be a way to show reverence for nature, and that they have no intention of expressing disrespect. These groups have competing interpretations of the site and the act of climbing. But such disagreements can be managed, if not entirely resolved. For example, these groups might compromise: The climbers might agree not to climb the Tower during the month of June when the tribes are conducting ceremonies around the monument. The Park Service could also discuss the Tower's spiritual significance on their website and post interpretive signs at the monument asking visitors to treat it with respect. This might be enough to recognize and honor the differing ways these groups value the landscape.

Unfortunately, however, such resolutions are not always possible. In our case, Judy K. may have to disturb the rock piles that might be burial mounds in order to protect the forest ecosystem by removing the diseased black oaks. The question is, would such an act be unjust, and if so, why?

To answer this question, it's useful to turn to philosophers who deal with theories of identity and recognition, such as Iris Young and Nancy Fraser. These theorists argue that lack respect or recognition in the form of insults, degradation, and devaluation is a harm and can be considered an injustice. When a social group suffers from lack of respect or simple non-recognition in the public realm — when they are characterized as less than human, their accomplishments devalued or denied, and their aspirations treated as unimportant — they suffer an injury that affects their ability to participate in politics and other important social activities. These disrespectful actions can exclude a social group from the community. Other citizens are less likely to consider their interests or point of view, and may treat them as outsiders or even dangerous invaders. Such disrespect can lead to other kinds of injustice: They might be denied economic opportunities, targeted by law enforcement, or discouraged from voting, for example. We're unlikely to achieve distributive and procedural justice if some people are treated as social pariahs or nonpersons. Justice requires treating equals equally, and this generally includes respecting the cultural, ethnic, racial, religious and other important elements of personal identity.

Of course, it's easy to agree that justice, not to mention basic civility, requires us to treat people with respect. The hard part is deciding what counts as disrespect or misrecognition. We're dealing here with symbolic actions, which are always open to interpretation. We should acknowledge the contributions that different groups have made to our national heritage. But how much acknowledgement is enough? For example, Yosemite National Park was established and protected in its early years by African American buffalo soldiers. That's certainly an interesting piece of history, but is it an injustice to African Americans if park rangers fail to mention it? What about Judy K.'s problem: Is it disrespectful to disturb burial mounds in order to protect the forest? It is important here to note that the Potawatomi people share the history of dispossession that most Native Americans experienced as the American government asserted its control over the land. Most of the Potawatomi people were forced, often at gunpoint, to relocate to Oklahoma. This history of injustice cannot now be changed, but it provides part of the context in which Judy must make decisions about the land. For example, she might want to consult with the Potawatomi tribe before making any decisions. This land is clearly part of their history, and acknowledging that would be a way to show them the respect they deserve.

The duty to show respect to human social groups leads to a further consideration: Do we have a duty to show respect to nature itself? Recall that some theorists (like Murray Bookchin) believe that our environmental problems stem from our domination and unjust exploitation of the natural world. If that theory is persuasive, then developing an ethically sound relationship to the natural world should involve showing respect for nature. As discussed in Sect. 4.2.3, philosopher Paul Taylor offers a set of principles and practices that would demonstrate such respect, including setting aside nature preserves, being careful not to cause damage to existing ecosystems without a very good reason, and simply leaving natural phenomena and processes alone. We can justify wilderness preservation on these grounds.

But it's not clear how to apply the principle of noninterference to a highly managed landscape like Spring Lakes. (Taylor himself limits his discussion of respect for nature to wild landscapes.) Judy K. can't simply refuse to interfere with natural processes on her property. Human interference has been occurring for generations. Indeed, the entire landscape is the result of perhaps hundreds of years of human management. And in a world affected by anthropogenic climate change, the same is rapidly becoming true of every ecosystem: *All* landscapes are increasingly the joint creation of human and nonhuman forces. What constitutes "respecting nature" in this world is a difficult but important question — and not one that philosophers alone can answer. Showing respect for nature is ultimately a matter of symbolic action. Respect must find expression in cultural practices, and that is generally the domain of artists, religious leaders, and other culture workers. And this is another reason why environmental education should include training in the arts and humanities.

7.7 The Problem of the Diseased Oaks Revisited

In sum, there may be no easy or obvious answer to Judy K.'s dilemma about the oaks and the burial mounds. It should be clear by now that the problem cannot be solved simply by drawing on ecological knowledge. Landscapes are storied; they have history and are invested with many kinds of meaning. And everything we do to and on the land potentially adds to that meaning. Judy is therefore not just managing a business and an ecosystem; she is managing meaning. She needs the imagination, sensitivity, and creativity of an artist.

But, like managing a business and an ecosystem, creating and managing the meaning of a place is inescapably a collective task. Everyone who lives on or has an interest in Spring Lakes will contribute to its stories. The story of humans in this place began long before European settlement, with the Potawatomi people who called this land their home. Others owned the land before Judy's family came into possession, and the land is part of their history as well. Tenants have been married there; pets were buried there; the ashes of former residents scattered there. Gardens were planted, tended, forgotten, and rediscovered. Adventures were planned and prosecuted. Many children had their first encounters with the mysterious, humble, curious, slimy, furry, scary, glorious, and ineffable aspects of nature there. The nonhuman members of the biotic community also have stories — the muskrats, herons, raccoons, and deer, and even the forests, marshes, and ponds can be considered characters in this unfolding tale. Judy is only one of many authors writing the story of Spring Lakes. Her decisions must be taken with careful attention to these other voices and perspectives, all of whom will have something to say about what came before and where the story should go next.

Further Reading

Brennan, A., Lo, Y.S.: Understanding Environmental Philosophy. Acumen Publishing, Durham (2010)

Callicott, B.: In Defense of the Land Ethic. SUNY Press, Albany (1989)

Cronon, W.: The trouble with wilderness. In: Cronon, W. (ed.) Uncommon Ground: Toward Reinventing Nature. WW Norton, New York (1995)

Krieger, M.: What's wrong with plastic trees? Science. **179**, 446–55 (1973)

Leopold, A., Sand County, A.: Almanac with Essays on Conservation from Round River. Ballantine Books, New York (1966)

McKibben, B.: The End of Nature. Anchor Books, New York (1990)

Mill, J.S.: On nature. In: Three Essays on Religion. H. Holt, New York (1874)

Saito, Y.: Appreciating nature on its own terms. In: Carlson, A., Lintott, S. (eds.) Nature, Aesthetics, and Environmentalism: From Beauty to Duty. Columbia University Press, New York (2008)

Strayed, C.: Wild: From Lost to Found on the Pacific Crest Trail. Alfred A. Knopf, New York (2012)

Taylor, P.: Respect for Nature. Princeton University Press, Princeton (1986)

Young, I.: Justice and the Politics of Difference. Princeton University Press, Princeton (1990)

Chapter 8
Stewardship as a Vocation

Contents

8.1 The Problem of Stewarding Spring Lakes

One of the things that worries Judy is what will happen to Spring Lakes when she's gone. We discussed in Chap. 5 the challenges of managing the land in the interests of future generations, but that discussion assumed that there would *be* some future land managers who would care about its history. As Judy grows older, she has to confront the question of who those future land stewards will be. She has four children, but do any of them want Spring Lakes? The property requires a good deal of hands-on management. Whoever takes on this responsibility will have to live there, far from the amenities and opportunities of major cities. Will any of her children be willing to dedicate a substantial part of their life to this sort of stewardship? It might seem to them like a sacrifice, to remain tied to a demanding job in this small corner of Michigan.

Judy herself didn't plan to become a land steward. Her career choice was education; she has spent most of her life working in the public school systems. But when she and her husband bought this property in 1970, the land and buildings were in rough condition. The place needed a lot of work. So, like many people, Judy found herself getting involved in land stewardship organically, without having given a great deal of thought to how it would fit into her life or what it would require of her. Her children are in a different situation. They grew up at Spring Lakes, but now they've made lives for themselves elsewhere. They would have to choose to come

K. K. Smith, *Exploring Environmental Ethics*,
AESS Interdisciplinary Environmental Studies and Sciences Series,
https://doi.org/10.1007/978-3-319-77395-7_8

back and take care of Spring Lakes. Choosing such a life is a complex matter, affecting one's livelihood, family, and a host of other important dimensions of one's life. How does one approach a choice like this, a choice of vocation? Can and should environmental ethics inform such a choice?

8.2 The Concept of Vocation

Let's begin with fundamentals: What is a vocation and how should one choose one? The concept of vocation or calling is rooted in Christian theology. Originally it referred to a divine calling from God to an individual to a particular way of life: the ministry or priesthood, or a life of celibacy or marriage. Protestants expanded this concept of a calling, suggesting that God might call one to perform any sort of useful work: teaching, medicine, farming, or making hats might be one's calling. You could discover your calling through a process of introspection, by discovering where your talents lie and what sort of work makes you feel fulfilled, bringing you into harmony with others and with God. The concept of vocation gave meaning, even sanctity, to work: Your life's work could be seen as a part of God's plan and a way to honor Him.

Contemporary philosopher Wendell Berry believes the concept of vocation can be meaningful even without this theological context. If we choose our work with careful consideration of its ethical significance, if we see our work as a form of service to higher ideals and a way to come into harmony with the world, then we may find greater meaning in work (and we may rely less heavily on consumption to give our lives meaning). "Work is necessary to us," he tells us, "as much a part of our condition as mortality; … good work is our salvation and our joy; …shoddy or dishonest or self-serving work is our curse and our doom" (Berry 1977, p. 14) Berry's hope is that people seeking a meaningful vocation would not choose a job that involves exploiting people or places; they would choose a life of service and stewardship.

Berry is not suggesting that everyone should become a land steward as a career, however. Although he finds a great deal of meaning in farming, not everyone will share that vocation. His point is that stewardship can be thought of as a general, fundamental vocation, no matter what your job is. That is, protecting the planet is everyone's basic mission in life; we just have different ways of doing that (just as Christians argue that serving God is everyone's vocation, even though everyone performs a different part of that general task.) Whether you are a teacher, parent, salesperson, artist, athlete, investment banker, student, doctor, or insurance salesman, you can use your knowledge, skills, and resources to help protect the earth. In other words, Berry offers stewardship as a larger ethical framework that gives meaning to our choices of career and lifestyle. He challenges us to see how our work, hobbies, relationships, and other practices can contribute to our fundamental mission: to preserve the ecological basis of a good human life on the planet.

8.3 Consumerism

Most environmentalists focus not on work but on consumption as the main driver of environmental degradation. More properly, they focus on consumerism: the cultural orientation that leads people to find meaning, contentment, and acceptance through what they consume. Consumerism helps to explain why consumption keeps rising in affluent countries that have more than enough wealth to satisfy basic needs. But consumerism is intimately connected to how we work. If we don't find meaning in our work, and if our work occupies most of our free time, we may have to turn to consumption was a way to express ourselves and find an identity and meaning. Moreover, consumerism is driven largely by business interests: by the basic need shared by most business enterprises to increase sales and therefore ensure a steady income stream. This is the main purpose of advertising, which is a major promoter of consumerism. Government officials worried about economic growth may also promote it, and much of the entertainment industry also embraces these consumer values. Indeed, consumerism has even crept into schools and religious institutions.

Unfortunately, merely critiquing consumerism is unlikely to lead to cultural change. Consumerism must be challenged by a competing value system, something even more compelling and meaningful. That's why I have focused here on work and the concept of vocation. Finding meaning in other parts of one's life, including the work of taking care of the planet, is one way to gradually diminish the force of consumerism in one's own life. One can also explore how to find meaning in relationships, community service, avocations and spiritual practices. Consumerism will not and probably should not disappear; humans must consume to live, and it's proper to find meaning in making good consumer choices. (Recall our discussion of Hegel's theory of property in Sect. 6.2.1: Taking control of material resources is an important way that we express our identity.) But the critics of consumerism have a good point: Consumption should not be the only or primary source of meaning in our lives. Ultimately consumption should be a part of a larger framework of meaning, a sense of purpose that informs your work, your play, and your whole life.

8.4 Environmental Ethics on the Job

Of course, for most people, choosing a vocation means choosing a job, and most of us probably don't give much thought to environmental ethics when we make that choice. In choosing a job or career, we think about matching our skills to the job opportunities available. We might think about where we want to live or what larger goals we want to achieve, like raising a family, starting a business, combating global hunger, or writing the next great American novel. Environmental values may not seem relevant to our life goals. Except for a few committed environmentalists, most of us probably think of protecting the environment as something we do after work, when we have some extra time.

But there are good reasons for us to think about our environmental values when choosing a career, even if protecting the environment isn't how you make a living. First, if you work outside the home (and most American adults do so at some point in their lives), most of your productive time will be spent at work. How your workday is structured will determine how environmentally sensitive you can be. If you work long hours, you may have little time or energy for recycling, buying locally, or finding ways to reduce your electricity consumption. (Supporting this claim is an interesting 2006 study by David Rosnick and Mark Weisbrot, showing that people who work long hours tend to make lifestyle choices that are more resource-intensive: They eat our more often, have larger homes, and use more carbon-intensive forms of travel, for example.) If your job requires you to travel a lot, you will travel, regardless of the carbon footprint it leaves. If your job requires you to sell consumer goods, you'll be promoting consumerism, whether you like it or not.

Moreover, your job usually constitutes your primary social community, so the organization you work for will inevitably have some influence on your worldview and your values. For example, if you work for a company that creates genetically modified organisms (GMOs), you will probably come to see GMOs as a good thing, an important contribution to ending world hunger and creating a sustainable agriculture. That perspective will be reinforced by the corporate culture. Anyone who remained skeptical about GMOs might find themselves socially isolated in this workplace. It's hard to resist the influence of these forces on your values.

This suggests that in choosing a career, you are not only making a choice about how you want to relate to the natural world right now, you may be making a choice about how your preferences and values regarding nature will *evolve*. Philosophers and psychologists confirm that values change over the course of one's life. Much of this change is simply the result of getting older and having more experience; you can't predict all the ways your perspective on life will change over time. But we do often have some well-considered preferences about how our preferences will change. (Philosophers call these *second order preferences*.) For example, perhaps you currently like eating meat, but you're persuaded that a vegetarian diet is better for the planet. You would prefer to prefer a vegetarian diet. You can make choices now (learning how to make vegetarian dishes, hanging out with vegetarians, joining a food co-op that would expand the range of foods you're exposed to) that will help you develop such a preference.

Let me pause here to discuss these terms, "values" and "preferences." Values differ from preferences in that they are based on moral principles adopted after thoughtful consideration, so they are usually more stable than preferences. Liking ice cream is a preference; supporting humane treatment of livestock is a value. But values mean very little unless they are reflected in your interests, tastes, and habits. Recall Elizabeth Anderson's point that valuing something is not just an attitude; it involves conduct and expression, and your ability to value something depends on having a supportive social context that allows you to express that value. Therefore, as your social world and lifestyle changes, so will your values.

But you can influence the direction of your moral evolution by actively shaping your social world and controlling the kinds of choices you face. For example, if you want to reduce your carbon footprint, you may decide not to buy a car. You might instead find a place to live that has good public transportation, or where you can rely on walking or biking to get around. This is sometimes called *choice-editing*. Instead of deciding every morning whether to drive work, you limit your choices in advance, taking driving off the table. (This might also be considered a *pre-commitment strategy*: putting in place a mechanism that commits you to a course of action in advance.) You can also influence your moral growth through your choice of friends and entertainment and your involvement in social organizations.

All of this suggests that in choosing a job, one should consider not only how much it will pay but how much it will cost—in terms of time, relationships, and opportunities for living a good life.

8.5 Environmental Ethics in College

College is a good place to start thinking about how your workplace shapes your relationship to the natural world. Environmental Studies scholar Jim Farrell did just that in his 2010 book, *The Nature of College*. He set out to "uncover the intellectual and emotional patterns that connect us to the degradation of nature" in the life of a typical college student (Farrell 2010, p. xiii). College is a *totalizing* institution; it takes over one's entire life, precisely during a key period moral and intellectual development (from 18 to 22 years old). Farrell's point is that institutions like colleges structure the systems that determine what choices we have. College also offers conceptual frameworks for making sense of the world; it draws our attention to certain things (the world of scholarship, parties, and sports) and directs our attention away from others (where our electricity, water, and food come from, and how much they cost).

Farrell examines how college life teaches students how to be typical American consumers: dependent on cars (most college campuses are commuter campuses), tethered to computer and TV screens, having fun through drinking, partying, sex, and shopping. (There is now an entire marketing strategy called "back to school.") The average college student pays little attention to politics or religion; classes aren't cancelled on election day, nor are most students expected to attend religious services anymore. And students are protected from the ecological consequences of many of their choices. Usually they aren't billed for or even made aware of how much electricity or water they use, and they have very little choice over the source of their food. Most colleges hide these ecological connections instead of highlighting them.

While many students resist and try to become ecologically aware, they must work against powerful institutional structures and cultural forces. Farrell calls for a transformation of these structures, so that environmental values can be better engrained in the patterns of a student's everyday life, in how they work, play, eat, study, and socialize.

8.6 Moral Ecologies

Of course, it is naïve to think that we all have a lot of choices about what job we take or where we go to school. Most of us will have to take a less than ideal job just to pay the bills, or choose the school that accepts us and that we can afford. What can you do if you find yourself in an institution that doesn't accommodate your environmental values?

One obvious option is to work on changing the institution itself. This could involve encouraging more sustainable processes (reducing waste, recycling more, conserving energy), as well as pressing for greater flexibility in scheduling your work and classes. In other words, one could try to bring into one's workplace or school all of the strategies one uses at home to protect natural resources.

A second option is to develop a social network that will help support your values. A good deal of research on moral psychology shows that a person's moral choices are influenced by many factors, including one's basic personality, one's commitment to being a moral person, one's moral skills (that is, the knowledge and skills necessary to do the right thing), and one's social context. This social context is sometimes called a *moral ecology*. Ethicist Chuck Huff has done interesting research on what makes a person more or less likely to follow ethical principles in the workplace, and moral ecology is one of those factors. He writes, "Communities, their practices and their values, provide the menu of goods from which their members select those that fit into their personal narratives" (Huff et al. 2008, p. 286.) Huff studied whistleblowers — people who had risked their jobs in order to address ethical violations in the workplace — and others considered *moral exemplars* by their peers. He found that these people often tried to influence the moral ecology of their workplace, to leave work environments that they considered unsupportive of their values, and to find mentors and colleagues who shared their values (creating their own *microecologies*.)

It's hard to change one's personality or moral commitment, but you can develop moral skills and change the shape of your moral ecology. Specifically, you can find friends, mentors, and colleagues who share your values. These networks can help you identify ethical challenges and find strategies to deal with them. They may also become allies who will help you transform the moral ecology of your school or workplace.

8.7 Politics as a Vocation?

Transforming schools and workplaces could have a powerful effect on a society's overall sustainability. But an even more important venue for achieving social change is the political arena, from the local community to the national and international arenas. To be sure, politics does not often look like a good place to practice virtue of any kind. Politics involves conflict and competition, which doesn't always bring out the best in people. We have rich literature on how political power corrupts, from Machiavelli's *The Prince* to modern classics like Robert Penn Warren's *All the King's Men*. But politics is also an important arena for practicing virtue. It allows us to put our selfish interests aside and consider the public good. Working collaboratively in a diverse community requires us to develop social virtues, like tolerance and respect for others, compassion, humility, and patience. Even political competition can call for virtues like courage and self-sacrifice. And recall our discussion of ecological citizenship in Sect. 3.7.2: Our duty to our fellow citizens may require us to participate in politics, to make sure the government fulfills our collective responsibilities.

Ultimately, if we want to live good lives, we cannot afford to ignore the political arena. Our moral choices are shaped in fundamental ways by our social context. Whether we even have the freedom to make ethical choices depends on having a political system that protects such freedom. And if we expect to enjoy the benefits of that protection, surely we have a responsibility to help maintain and improve the system that provides it.

8.8 Stewarding Spring Lakes Revisited

Does this discussion help Judy K. and her family decide how to handle the generational transition, the passing of Spring Lakes from one generation to the next? It does offer some reasons for choosing to take care of Spring Lakes, reasons that some of her children might find compelling. I hope that our exploration of environmental ethics has shown how a life devoted to managing the land can be a good, choiceworthy life; it is one way to live a good human life in relationship to the natural world.

But it may be that none of Judy's children have the vocation for managing the property. After all, there are many other ways to cultivate a good relationship to the natural world. In that case, perhaps Judy should look beyond her immediate family for land stewards. For example, many states allow landowners to protect the natural values of their land by donating or selling it to an organization that will conserve it as a park, wildlife refuge, or some other kind of natural area. Such organizations include nonprofit corporations (such as the Trust for Public Land) as well as the state or federal government. For many families, this form of conservation is a meaningful way to show proper appreciation for the aesthetic, historical, and ecological value of the landscape. However, it is important to note that these legal options are available only if the community of which Spring Lakes is a part shares the social ideal of stewardship. Judy's fellow citizens must support these conservation policies by allowing their tax dollars to be used in this fashion.

Alternatively, Judy might sell the land to someone who seems to value it as she does, or leave it to her children to sell if they wish. If her community values the stewardship ideal, there should be potential buyers who would appreciate the care that Judy put into the land. But again, this possibility depends on the values held by the larger community. We suggested in Chap. 5 that the idea that we're bequeathing a good world to the future helps to satisfy our need for a meaningful life, that it is one way to make our lives matter. We see now that our ability to make our lives meaningful in this way may depend in part on whether our community shares the values we are trying to achieve. This is another reason to believe that our efforts to live a good life in relationship to nature are, and must be, a *collective* project. The natural world is our common heritage, and our common legacy. We must work together to realize its best possibilities.

Further Reading

Berry, W.: The Unsettling of America. Sierra Club Books, San Francisco (1977)

Farrell, J.: The Nature of College. Milkweed Editions, Minneapolis (2010)

Huff, C., Barnard, L., Frey, W.: Good computing: a pedagogically focused model of virtue in the practice of computing. Two parts. Journal of Information, Communication and Ethics in Society. **6**(3), 246–248; **6**(4), 284–316 (2008)

Rosnick, D., Weisbrot, M.: Are Shorter Work Hours Good for the Environment? Working Paper. Center for Economic and Policy Research, Washington, DC (2006)

Index

K. K. Smith, *Exploring Environmental Ethics*,
AESS Interdisciplinary Environmental Studies and Sciences Series,
https://doi.org/10.1007/978-3-319-77395-7

Printed in the United States
By Bookmasters